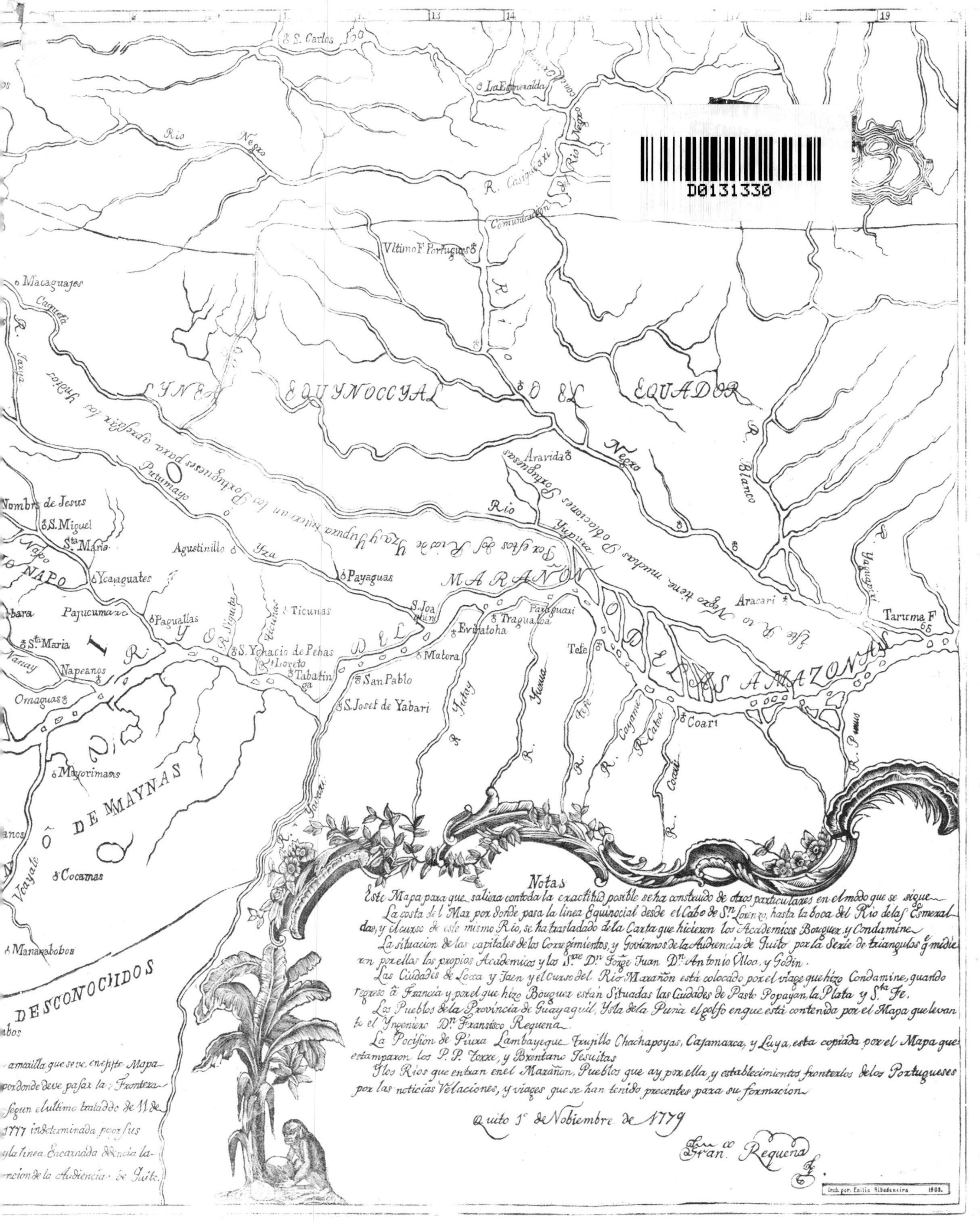
S. Carlos
La Esmeralda
Rio Negro
R. Casiquiari
Comunicacion
Ultimo F. Portugues
Macaguajes
Caqueta
R. Japura
LYNEA EQUYNOCCYAL DE EL EQUADOR
Aravida
Negro
R. Blanco
Putumayo
Rio Yupura
Por estos dos Rios de Yza y Yupura entran los Portugueses para apresar los Yndios
Este Rio Negro tiene muchas Poblaciones Portuguesas
Nombre de Jesus
S. Miguel
Sta. Maria
Napo
RIO NAPO
Agustinillo
Yza
Ycaraguates
Payaguas
MARAÑON O DE LAS AMAZONAS
Aracari
R. Yaguapixi
Taruma F.
Pajucumaro
Paguallas
Ticunas
S. Joaquin
Paraguaxi
Tragualoa
Evinatoha
Tefe
S. Yñacio de Pebas
Loreto
Tabatinga
Matora
San Pablo
S. Josef de Yabari
R. Jutay
R. Juxua
R. Cayame
R. Catua
Coari
R. Coari
R. Purus
Sta. Maria
Nanay
Napeanos
Omaguas
Mayorimasas
MAYNAS
Cocamas
R. Yavari
Ucayale
Manarabobos
DESCONOCHIDOS
Notas
Este Mapa para que saliera con toda la exactitud posible se ha construido de otros particulares en el modo que se sigue
La costa del Mar por donde pasa la linea Equinocial desde el Cabo de Sn. Lorenzo, hasta la boca del Rio de las Esmeraldas, y el curso de este mismo Rio, se ha trasladado de la Carta que hicieron los Academicos Bouguer, y Condamine
La situacion de las capitales de los Corregimientos, y Goviernos de la Audiencia de Quito por la Serie de triangulos q. midieron por ellas los propios Academicos y los Sres. Dn. Jorge Juan Dn. Antonio Ulloa, y Godin.
Las Ciudades de Loxa y Jaen y el Curso del Rio Marañon está colocado por el viage que hizo Condamine, quando regreso á Francia y por el que hizo Bouguer están situadas las Ciudades de Pasto Popayan, la Plata y Sta. Fe.
Los Pueblos de la Provincia de Guayaquil, Ysla de la Puna el golfo en que está contenida por el Mapa que levanto el Yngeniero Dn. Fransisco Requena
La Pocision de Piura Lambayeque Truxillo Chachapoyas, Cajamarca, y Luya, esta copiada por el Mapa que estamparon los P. P. Torre, y Brentano Jesuitas
Y los Rios que entran en el Marañon, Pueblos que ay por ella, y establecimientos fronterizos de los Portugueses por las noticias Velaciones, y viages que se han tenido presentes para su formacion
Quito 1º de Nobiembre de 1779
Fran. Requena
Grab. por. Emilia Ribadeneira 1908.

CRUDE

STORY
PABLO FAJARDO · SCRIPT
SOPHIE TARDY-JOUBERT · DRAWING AND COLOR
DAMIEN ROUDEAU

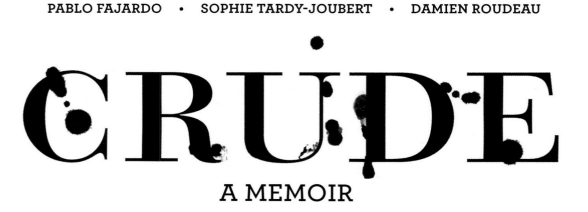

A MEMOIR

graphic mundi

Never Give Up

It's a universal story, of course, but what does it tell us? Claude Lévi-Strauss's famous book *Tristes Tropiques* begins with the sentence "I hate traveling and explorers." These words have been quoted a thousand times; nevertheless, they continue to haunt me. I now consider the intrusion into and exploration of unspoiled places to be a crime. Must we accept that everything will be destroyed? Must we accept that not a single inch of our shared Earth can escape exploitation?

The disaster described in this book is extraordinary, and its consequences for the Indians of Ecuador and the animals, plants, and spirits of the "Great Forest" have no discernible limits. This pointed insanity is enough to bring us to tears, while it delights those who caused it. That's because we no longer speak the same language. Yes, I'm claiming that we—those who prefer life and its enchantments, its millions of forms, its strange languages—have broken with these others. It pains me to write it, but we no longer belong to the same universe as they: the destroyers, the demolishers, the assassins of the world's beauty.

"Crude" is merely a metaphor—the allegory of a process that has completely escaped human control. The oil industry *has to* go on, because it *must* go on. Just like the industries for pesticides, plastics, fishing, and logging.

Is this sentiment defeatist? Pessimistic? I don't believe so. When we see the damage that can be inflicted on a country by a single corporation—and there are thousands like it—there remains only one future for those of us who are still standing: rebellion. Real rebellion, which requires risking everyone and everything. The choice has never seemed so clear: submit or revolt.

Certainly, Chevron, who acquired Texaco decades after it started extracting oil in Ecuador, is not alone in behaving this way. Shell transformed the Niger Delta—once a great hymn to ecological wealth—into a cesspool where the Ibo, Ogoni, and Ijaw peoples now struggle to survive. Wherever it gushes forth, oil inevitably causes widespread corruption and massive pollution, and this is made known to the public in perhaps just 1 percent of cases. Even Ecuador, under the anti-globalization leadership of Rafael Correa, became implicated, when in 2016 it began extracting oil on Waorani land, threatening the area's staggering 696 species of birds, 2,274 species of trees, and 169 species of mammals. Of course this was after the countries to the north refused to offer compensation for the non-exploitation of the oil reserves in Yasuni National Park.

Despite growing opposition, the French government, too, is getting involved, in its attempt to operate a gold mine in the middle of Guyana in a sacred and untouched piece of the magnificent Amazon rainforest. Will it succeed? We must stand together against the plundering of Ecuador: It is better to stop the damage before it occurs than to seek reparations later.

Above all, do not believe that the actions of Chevron/Texaco do not concern us here in France. Didn't Total, a transnational company, absorb the French company Elf in 2000? And with it, its network of influence, Françafrique, and the ever-changing array of puppet regimes in Africa that are happy to supply oil? Wasn't Total implicated in Burma, Iraq, Russia, Libya, Nigeria? Wasn't it responsible for the dreadful *Erika* oil spill?

Get ready, now, to enjoy this page-turner of a graphic novel. Make way for Pablo Fajardo, the tireless and resolute lawyer who fights, with all his authenticity and directness, for our cause. For your cause. For the cause of those of us who are now convinced that we must never give up. We must never give up again.

Fabrice Nicolino

French journalist and founder of the movement "Nous Voulons des Coquelicots," working to ban the use of synthetic pesticides in France

For the victims of oil pollution.
For Camille, and for all the places to defend.

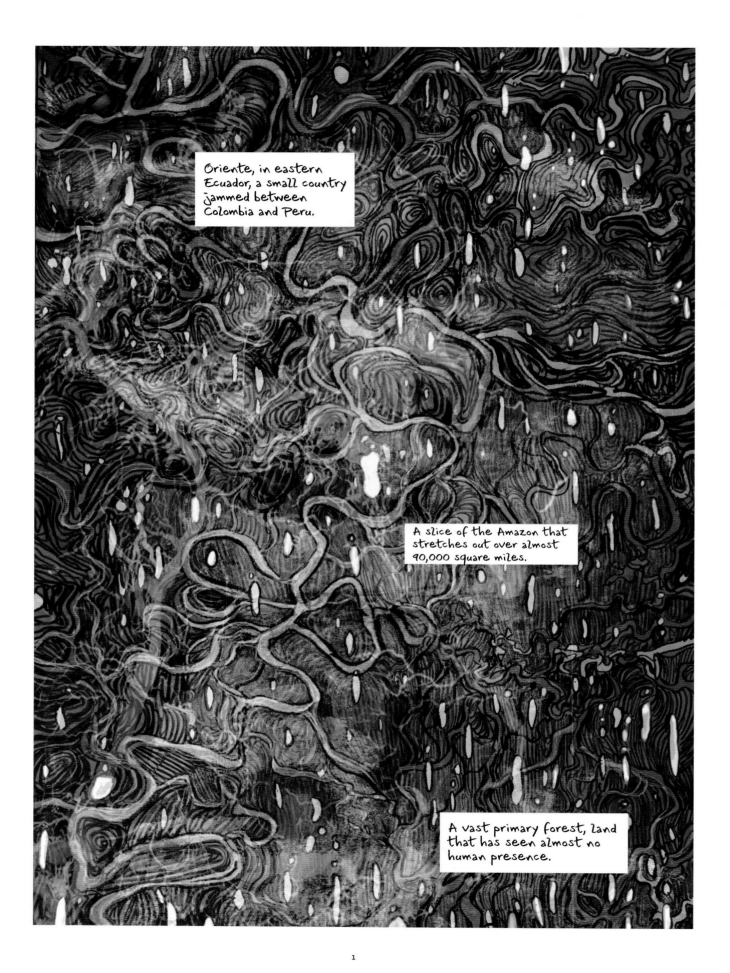

Oriente, in eastern Ecuador, a small country jammed between Colombia and Peru.

A slice of the Amazon that stretches out over almost 90,000 square miles.

A vast primary forest, land that has seen almost no human presence.

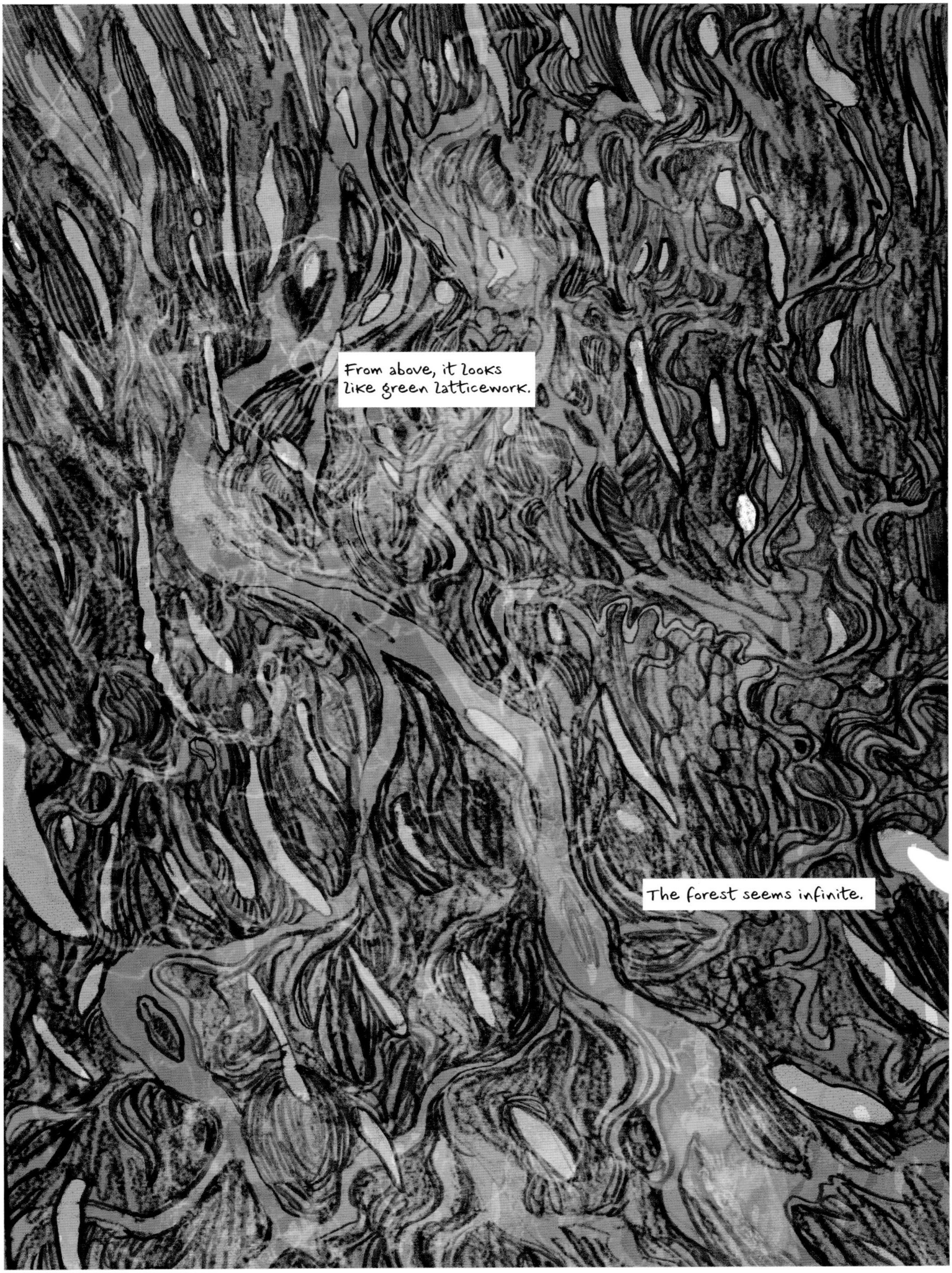
From above, it looks like green latticework.
The forest seems infinite.

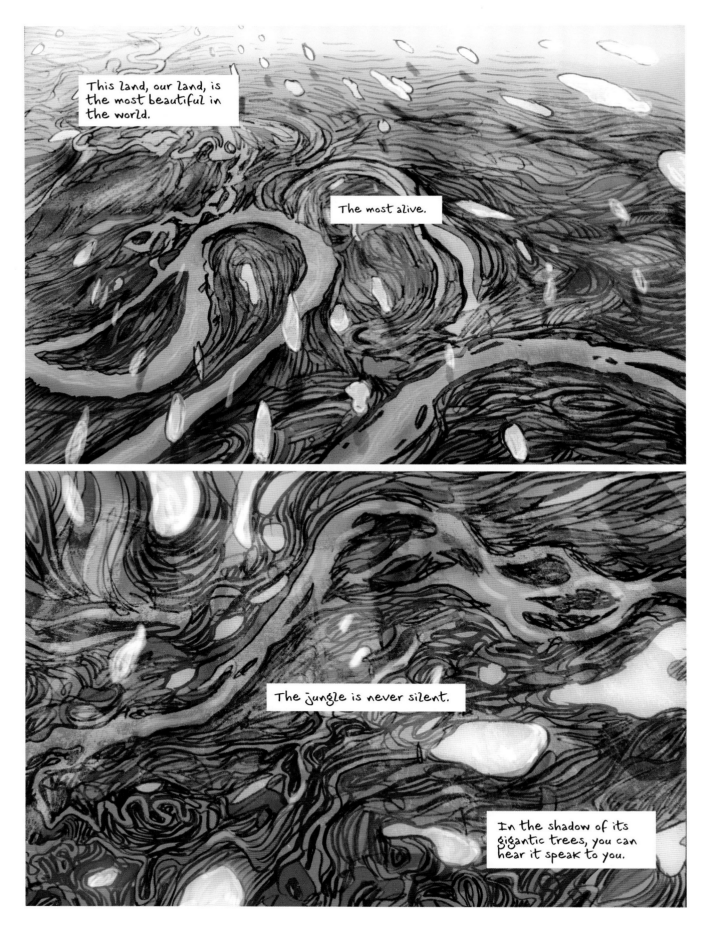

This land, our land, is the most beautiful in the world.

The most alive.

The jungle is never silent.

In the shadow of its gigantic trees, you can hear it speak to you.

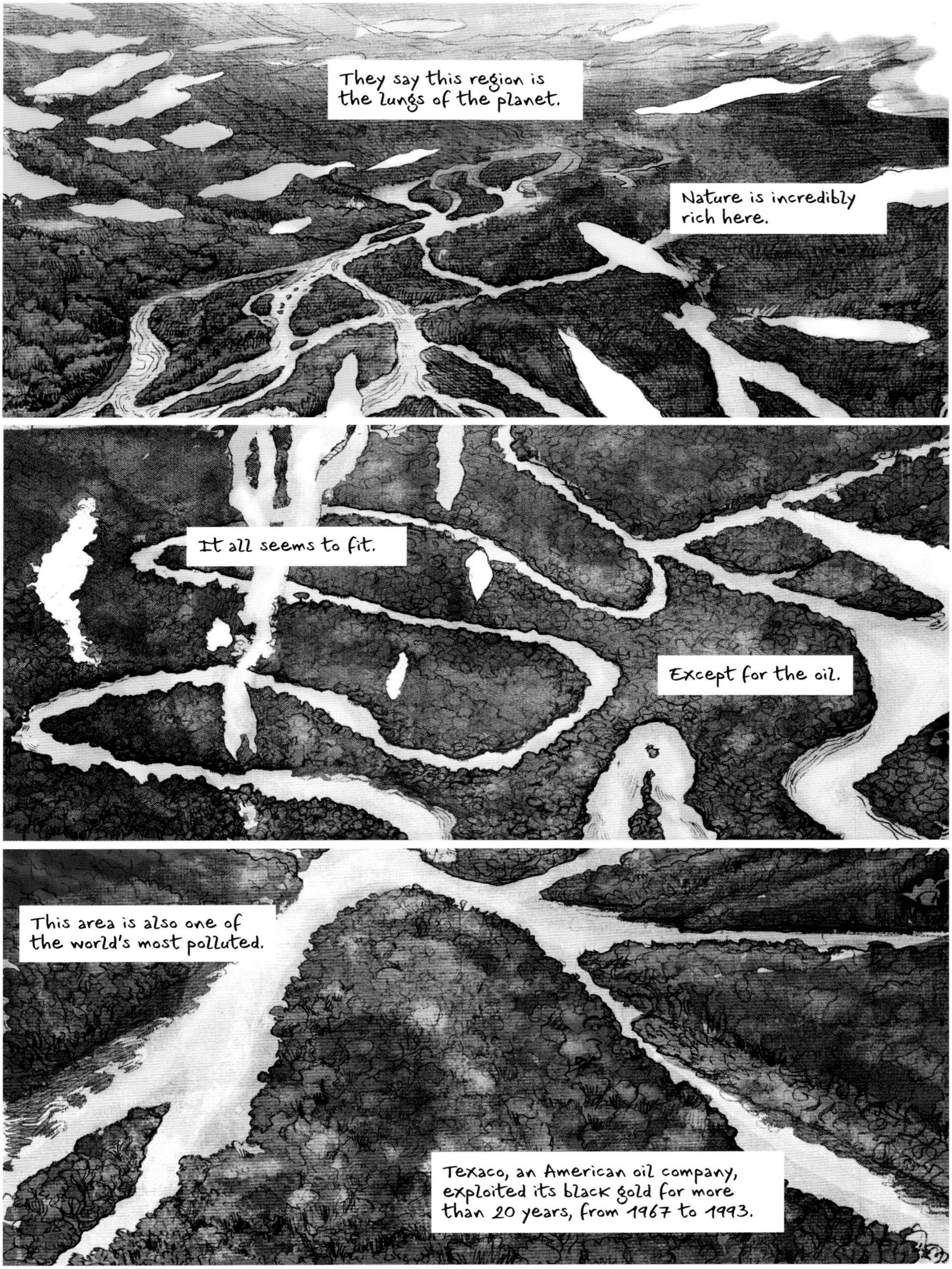
They say this region is the lungs of the planet.
Nature is incredibly rich here.
It all seems to fit.
Except for the oil.
This area is also one of the world's most polluted.
Texaco, an American oil company, exploited its black gold for more than 20 years, from 1967 to 1993.

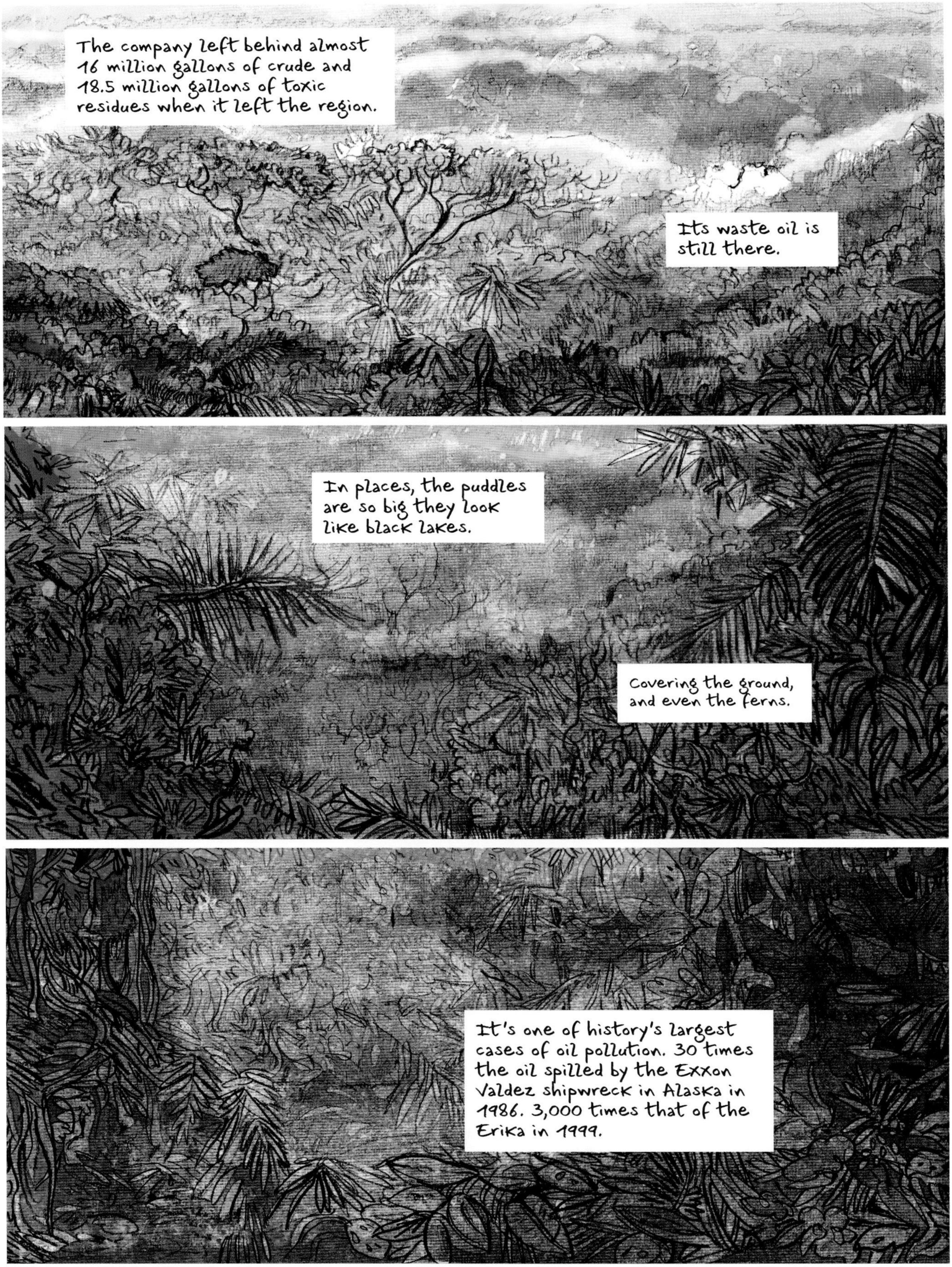
The company left behind almost 16 million gallons of crude and 18.5 million gallons of toxic residues when it left the region.
Its waste oil is still there.
In places, the puddles are so big they look like black lakes.
Covering the ground, and even the ferns.
It's one of history's largest cases of oil pollution. 30 times the oil spilled by the Exxon Valdez shipwreck in Alaska in 1986. 3,000 times that of the Erika in 1999.

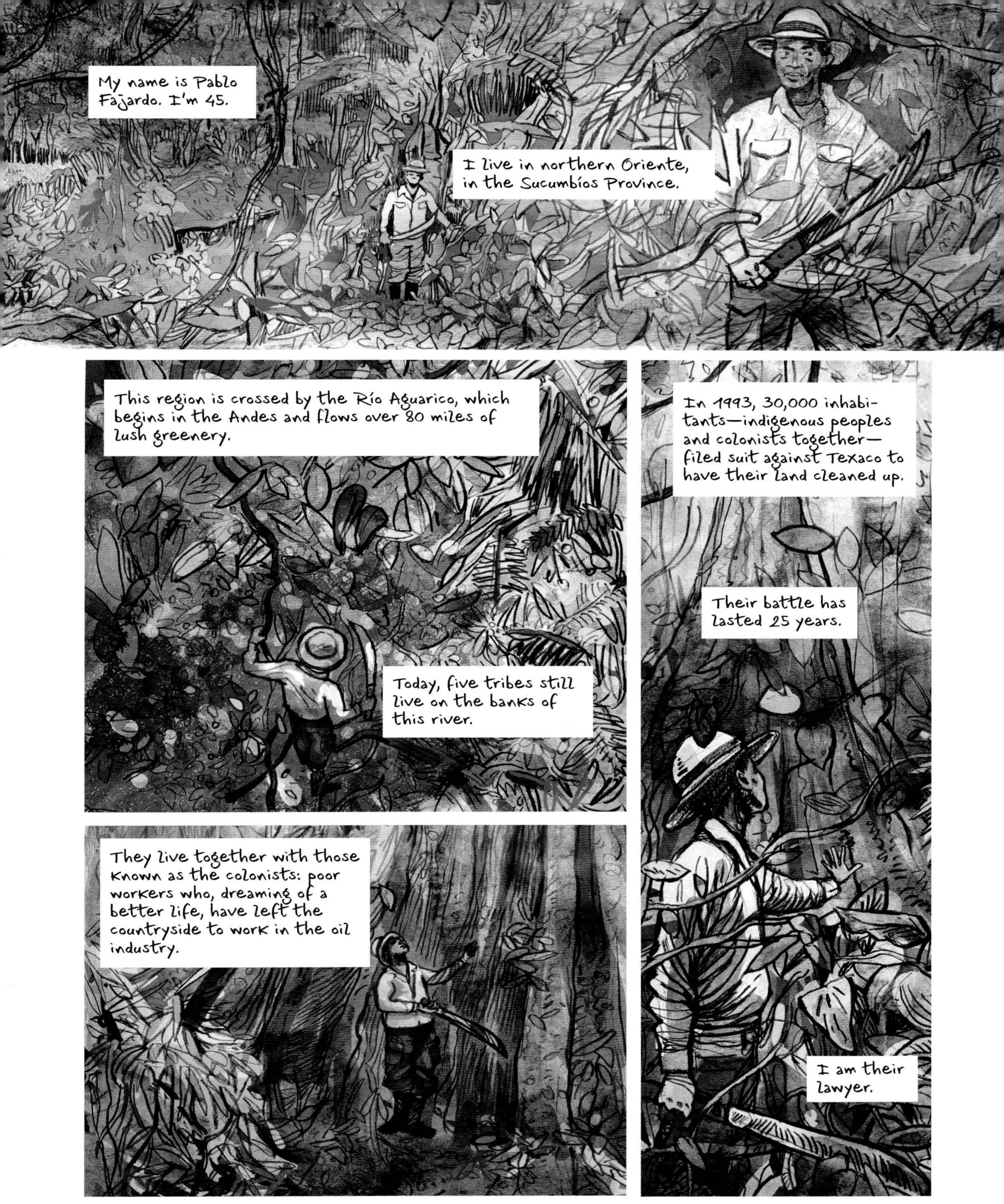
My name is Pablo Fajardo. I'm 45.
I live in northern Oriente, in the Sucumbíos Province.
This region is crossed by the Río Aguarico, which begins in the Andes and flows over 80 miles of lush greenery.
Today, five tribes still live on the banks of this river.
They live together with those known as the colonists: poor workers who, dreaming of a better life, have left the countryside to work in the oil industry.
In 1993, 30,000 inhabitants—indigenous peoples and colonists together—filed suit against Texaco to have their land cleaned up.
Their battle has lasted 25 years.
I am their lawyer.

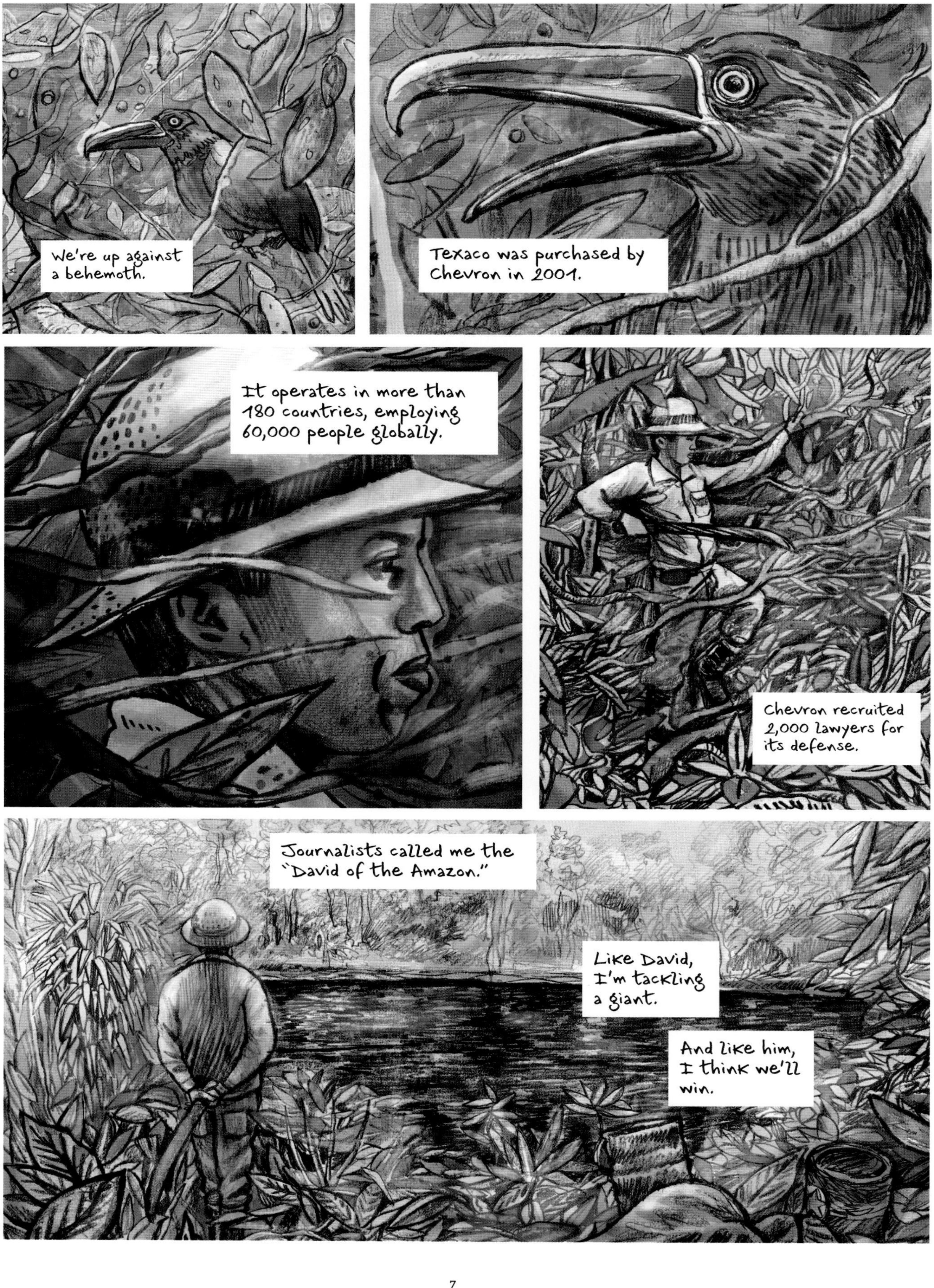
We're up against a behemoth.
Texaco was purchased by Chevron in 2001.
It operates in more than 180 countries, employing 60,000 people globally.
Chevron recruited 2,000 lawyers for its defense.
Journalists called me the "David of the Amazon."
Like David, I'm tackling a giant.
And like him, I think we'll win.

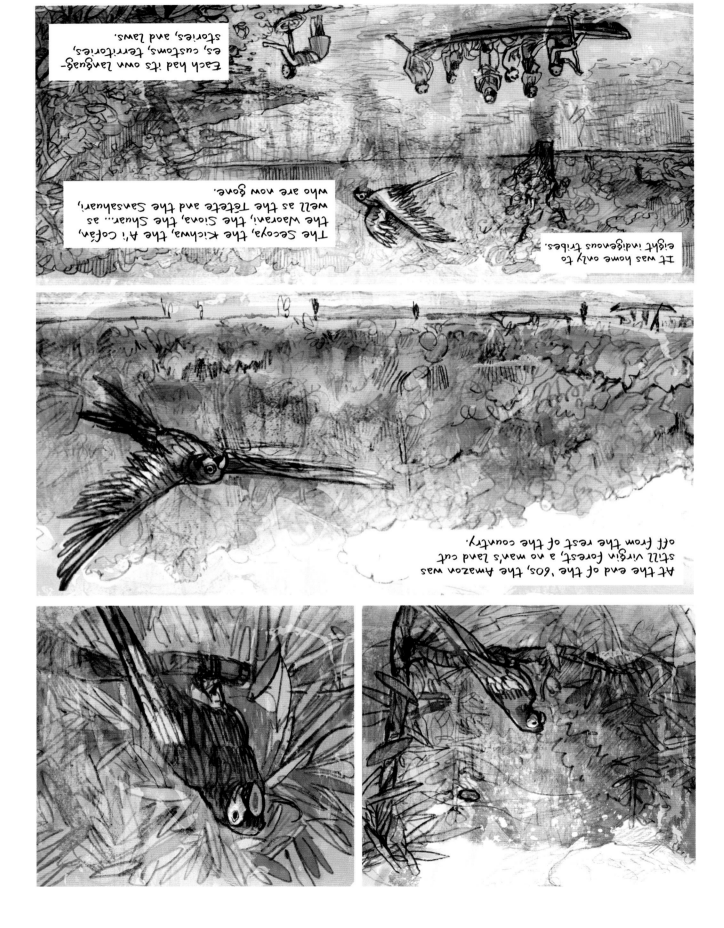

Each had its own languages, customs, territories, stories, and laws.

The Secoya, the Kichwa, the A'i Cofán, the Waorani, the Siona, the Shuar... as well as the Tétete and the Sansahuari, who are now gone.

It was home only to eight indigenous tribes.

At the end of the '60s, the Amazon was still virgin forest, a no man's land cut off from the rest of the country.

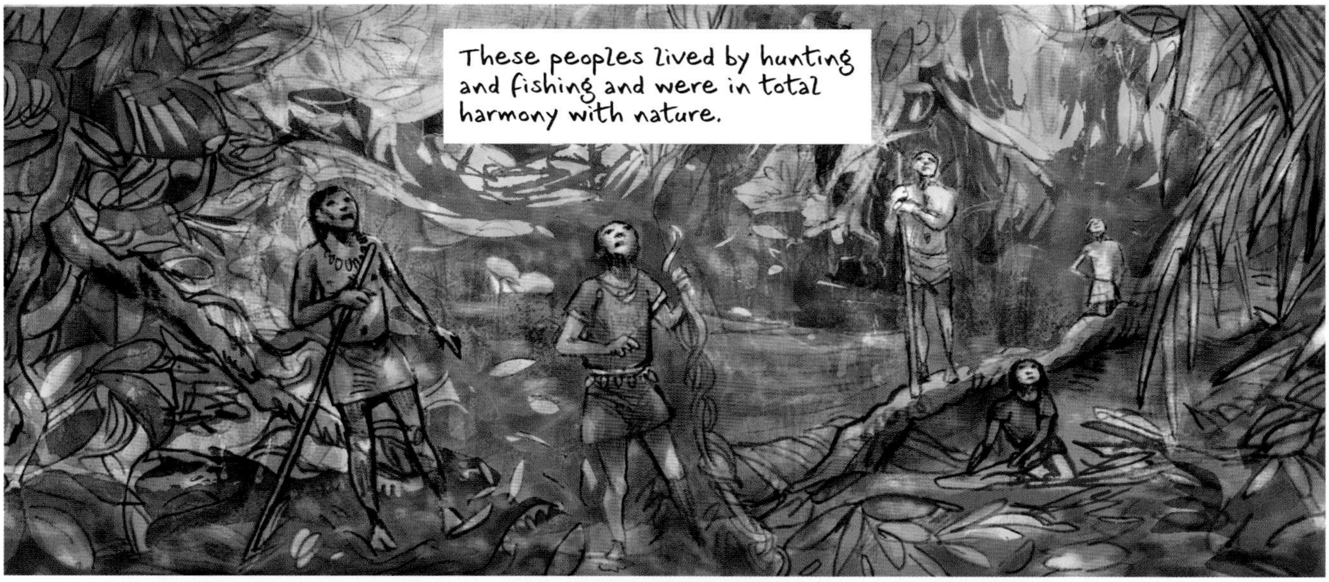
These peoples lived by hunting and fishing and were in total harmony with nature.

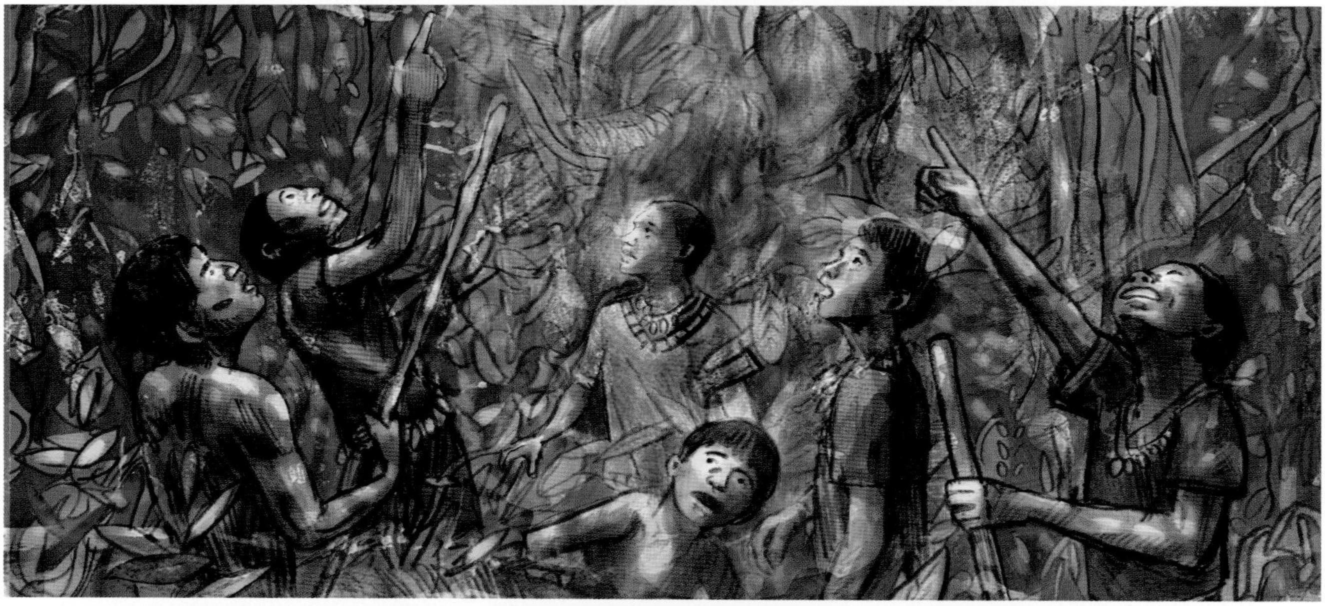

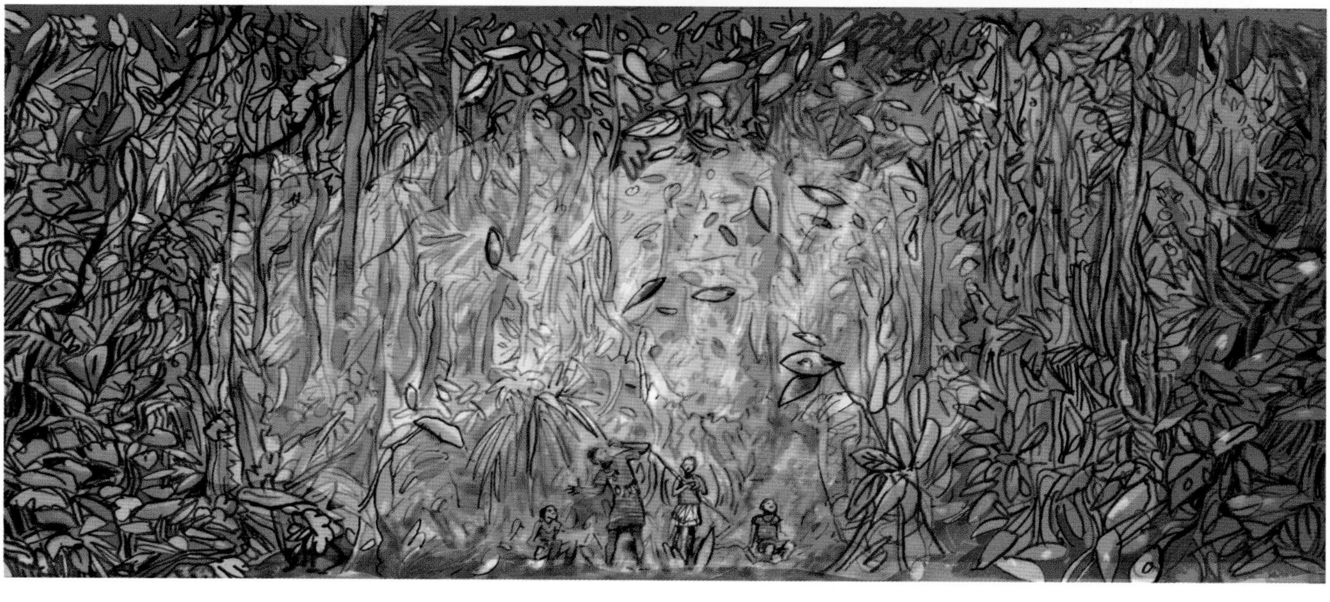

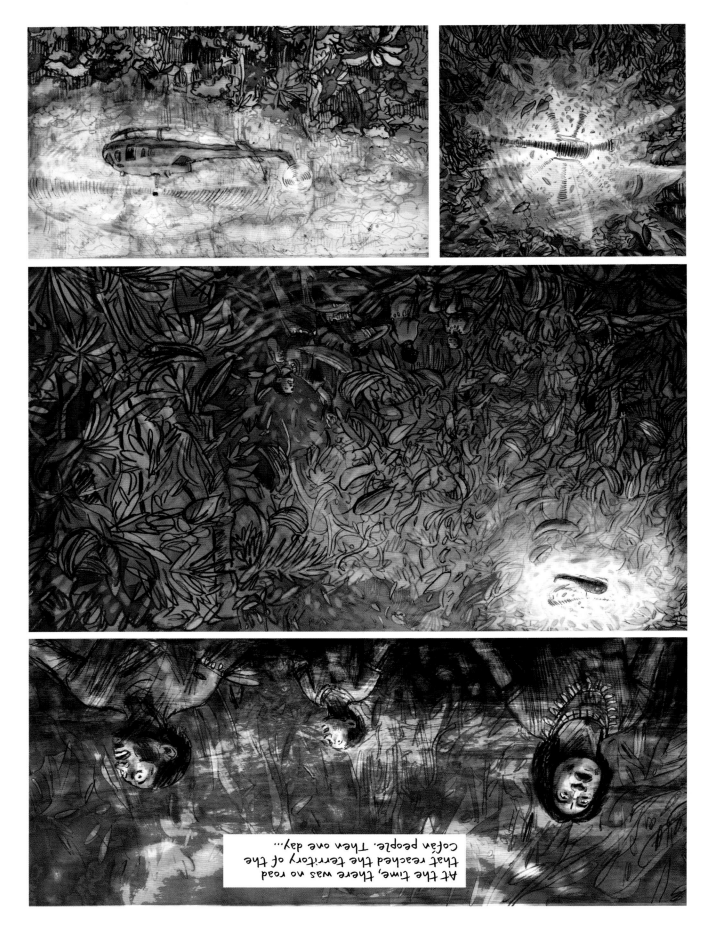

At the time, there was no road that reached the territory of the Cofán people. Then one day...

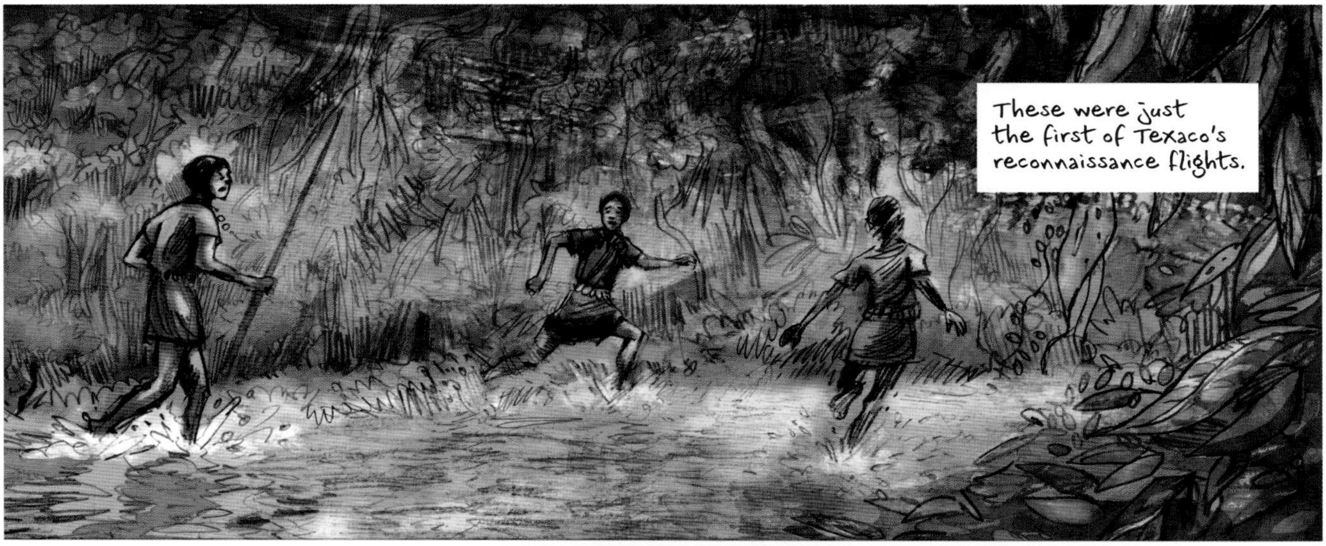
These were just the first of Texaco's reconnaissance flights.

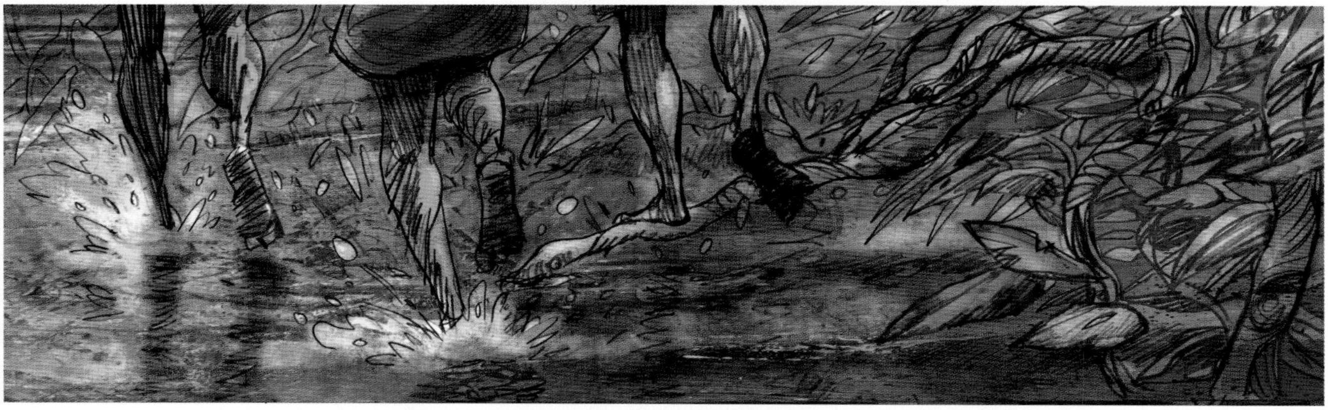

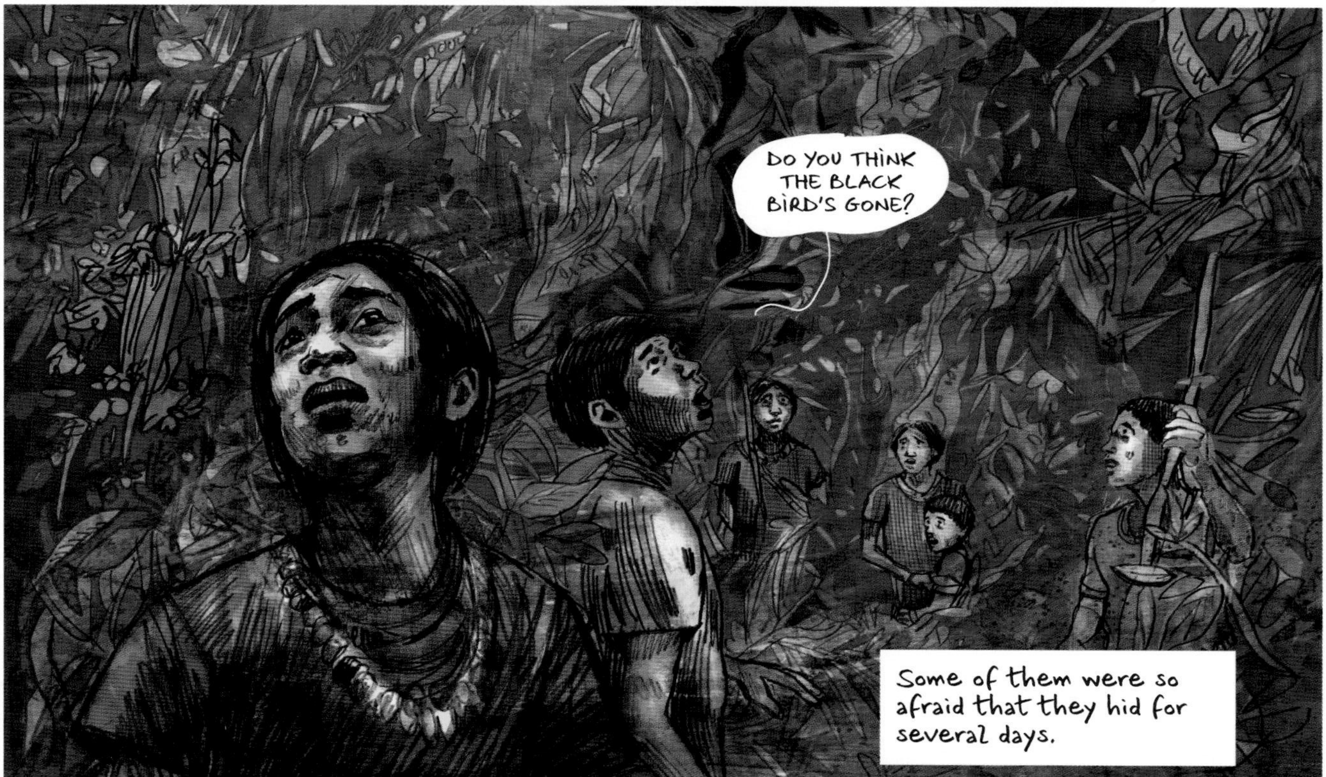
DO YOU THINK THE BLACK BIRD'S GONE?
Some of them were so afraid that they hid for several days.

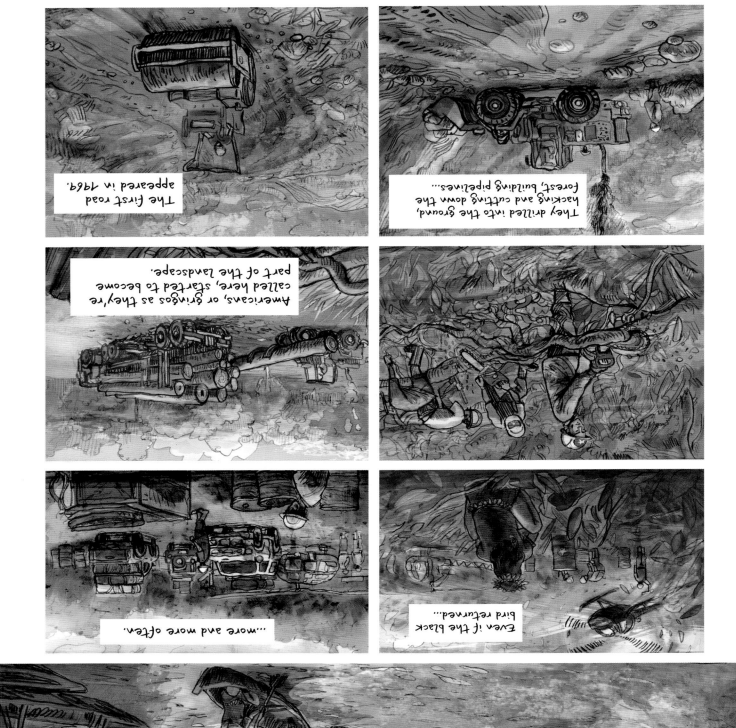

The first road appeared in 1964.

They drilled into the ground, hacking and cutting down the forest, building pipelines...

Americans, or gringos as they're called here, started to become part of the landscape.

"...more and more often.

Even if the black bird returned...

Life returned to normal.

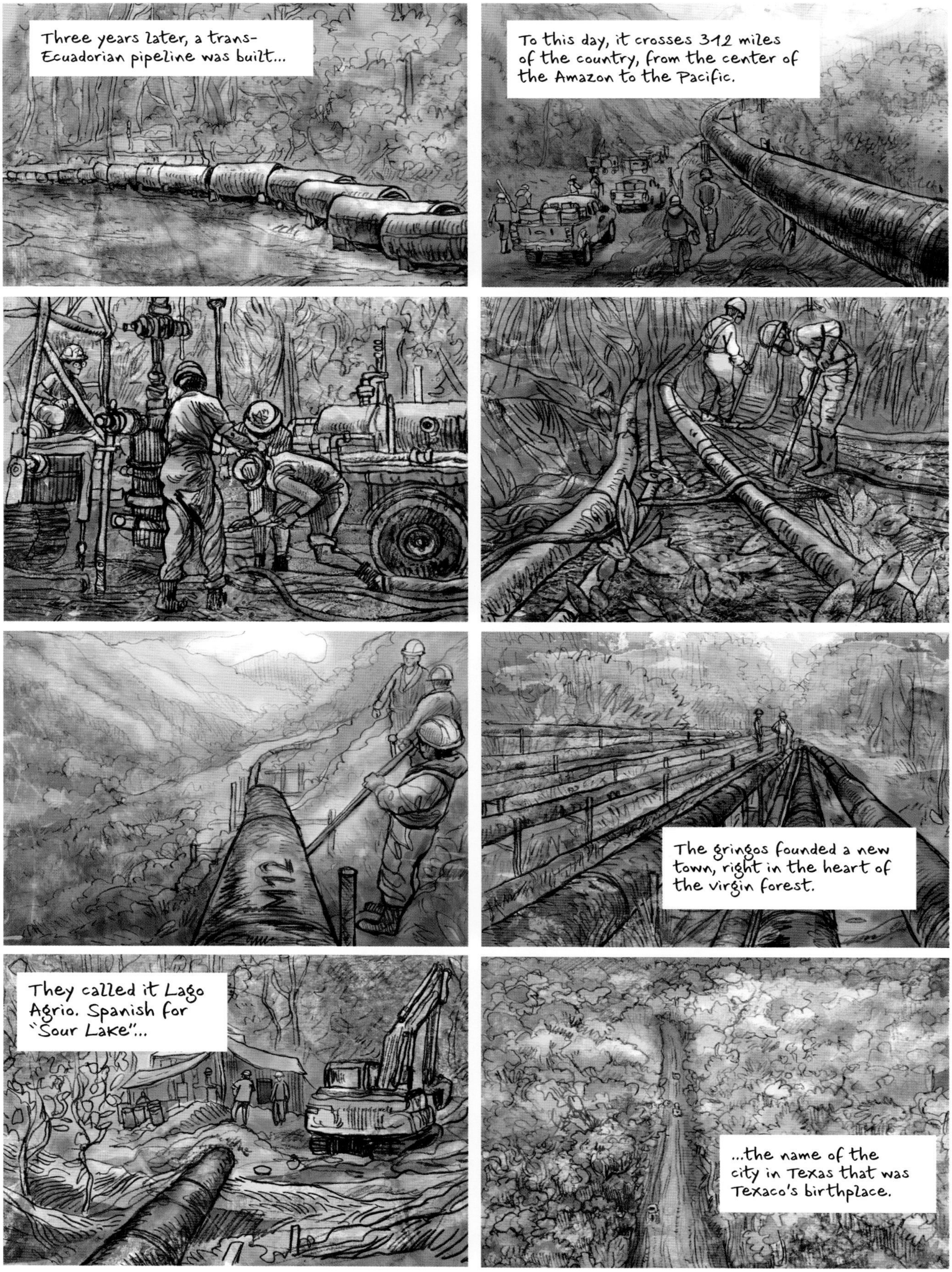
Three years later, a trans-Ecuadorian pipeline was built...
To this day, it crosses 312 miles of the country, from the center of the Amazon to the Pacific.
The gringos founded a new town, right in the heart of the virgin forest.
They called it Lago Agrio. Spanish for "Sour Lake"...
...the name of the city in Texas that was Texaco's birthplace.

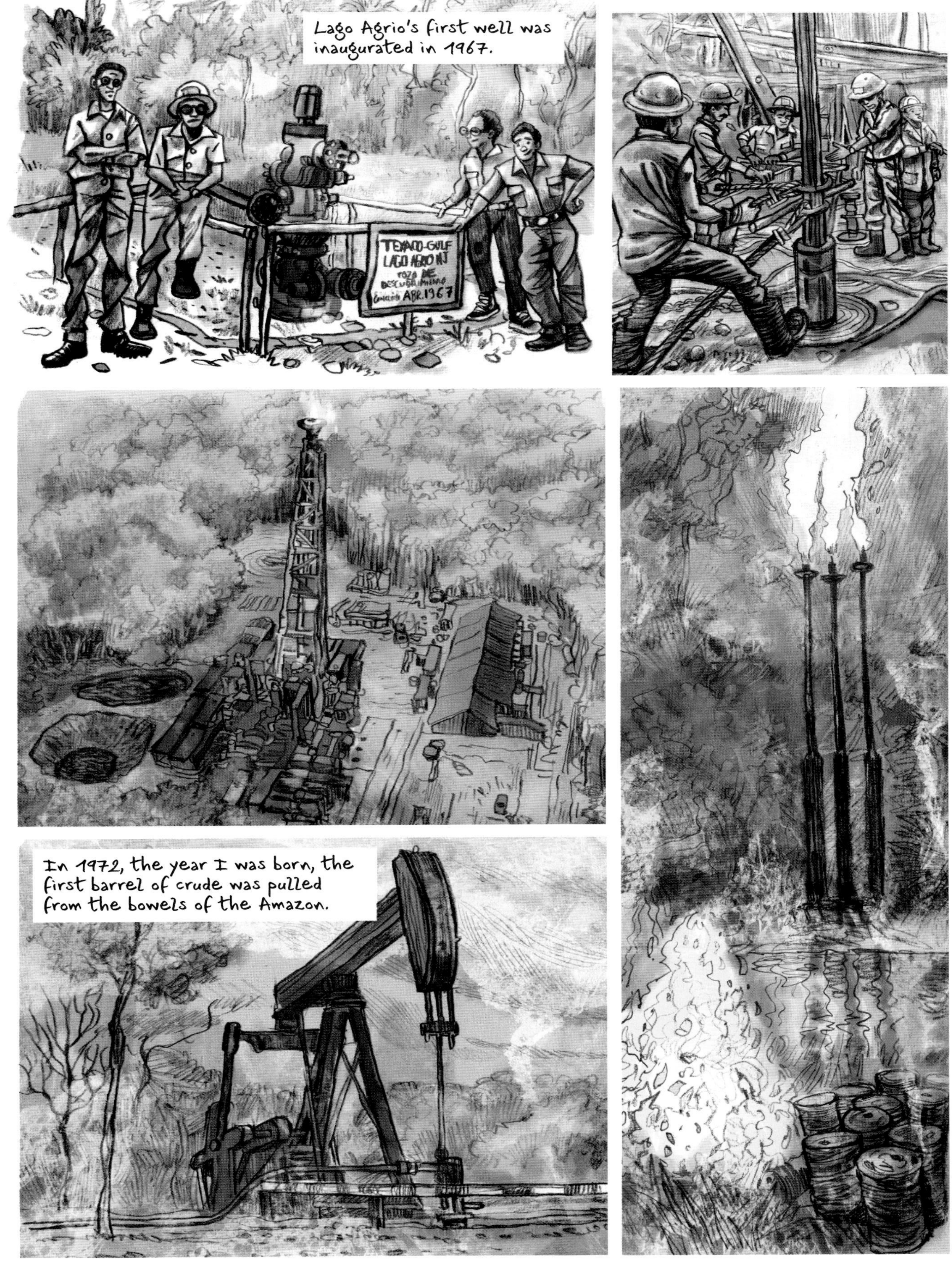
Lago Agrio's first well was inaugurated in 1967.
In 1972, the year I was born, the first barrel of crude was pulled from the bowels of the Amazon.

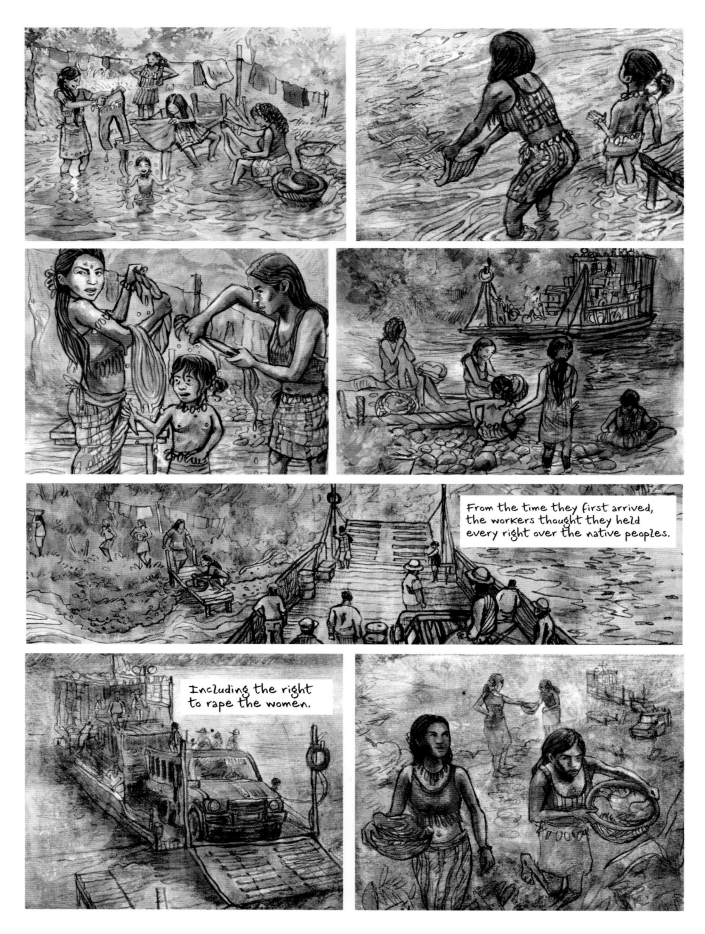

From the time they first arrived, the workers thought they held every right over the native peoples.

Including the right to rape the women.

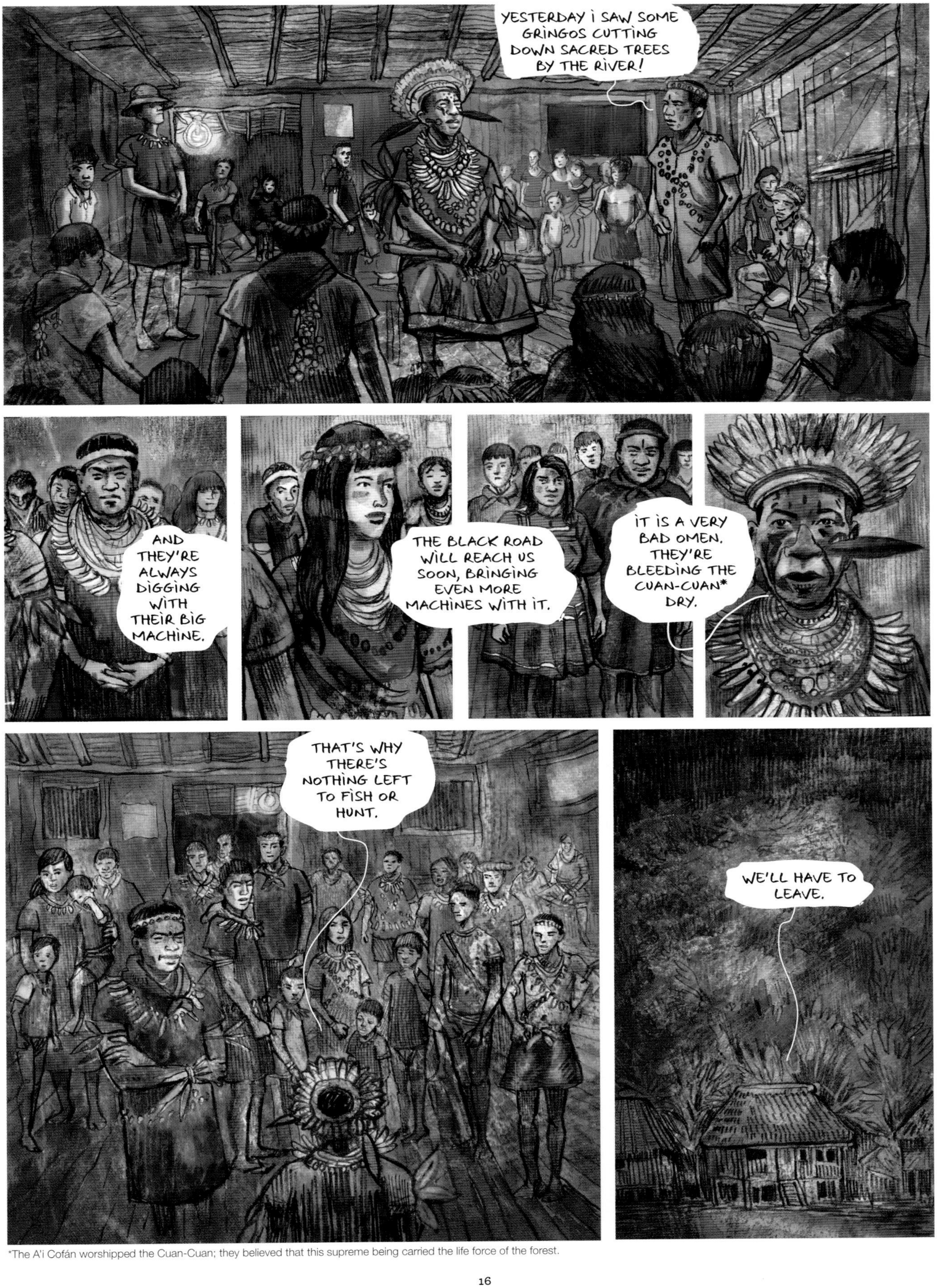
YESTERDAY I SAW SOME GRINGOS CUTTING DOWN SACRED TREES BY THE RIVER!
AND THEY'RE ALWAYS DIGGING WITH THEIR BIG MACHINE.
THE BLACK ROAD WILL REACH US SOON, BRINGING EVEN MORE MACHINES WITH IT.
IT IS A VERY BAD OMEN. THEY'RE BLEEDING THE CUAN-CUAN* DRY.
THAT'S WHY THERE'S NOTHING LEFT TO FISH OR HUNT.
WE'LL HAVE TO LEAVE.

*The A'i Cofán worshipped the Cuan-Cuan; they believed that this supreme being carried the life force of the forest.

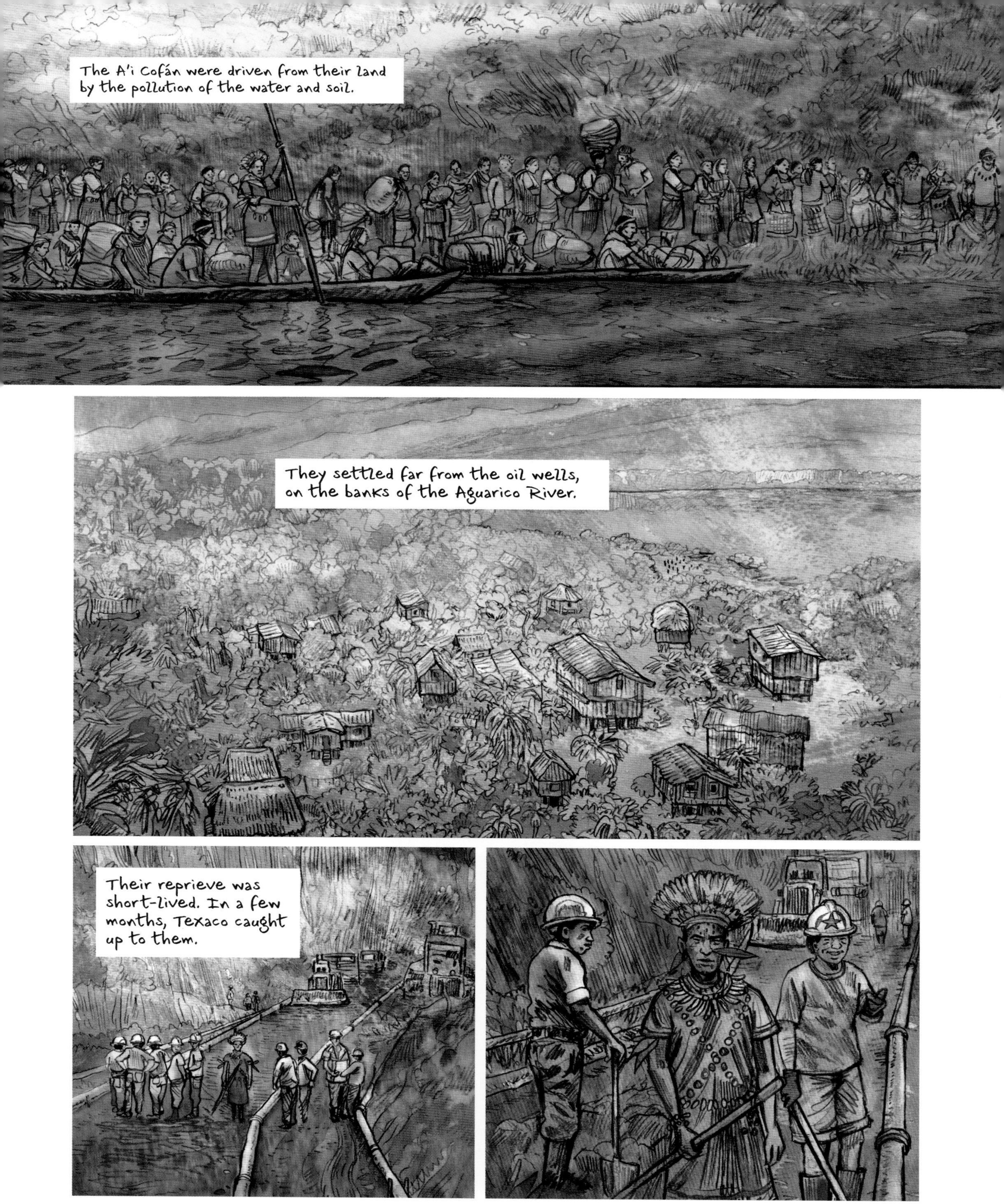
The A'i Cofán were driven from their land by the pollution of the water and soil.
They settled far from the oil wells, on the banks of the Aguarico River.
Their reprieve was short-lived. In a few months, Texaco caught up to them.

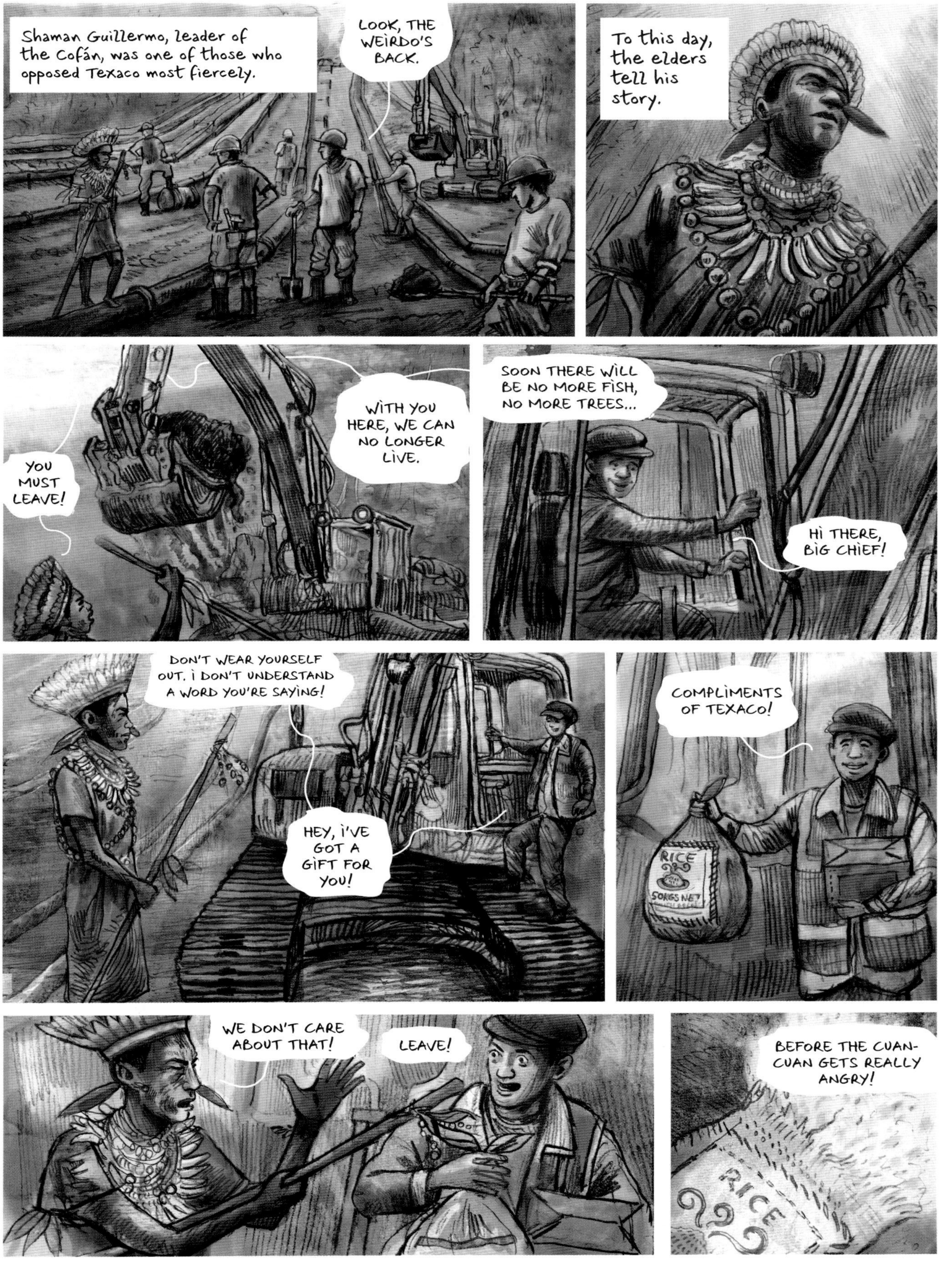
Shaman Guillermo, leader of the Cofán, was one of those who opposed Texaco most fiercely.
LOOK, THE WEIRDO'S BACK.
To this day, the elders tell his story.
YOU MUST LEAVE!
WITH YOU HERE, WE CAN NO LONGER LIVE.
SOON THERE WILL BE NO MORE FISH, NO MORE TREES...
HI THERE, BIG CHIEF!
DON'T WEAR YOURSELF OUT. I DON'T UNDERSTAND A WORD YOU'RE SAYING!
HEY, I'VE GOT A GIFT FOR YOU!
COMPLIMENTS OF TEXACO!
RICE
WE DON'T CARE ABOUT THAT!
LEAVE!
BEFORE THE CUAN-CUAN GETS REALLY ANGRY!
RICE

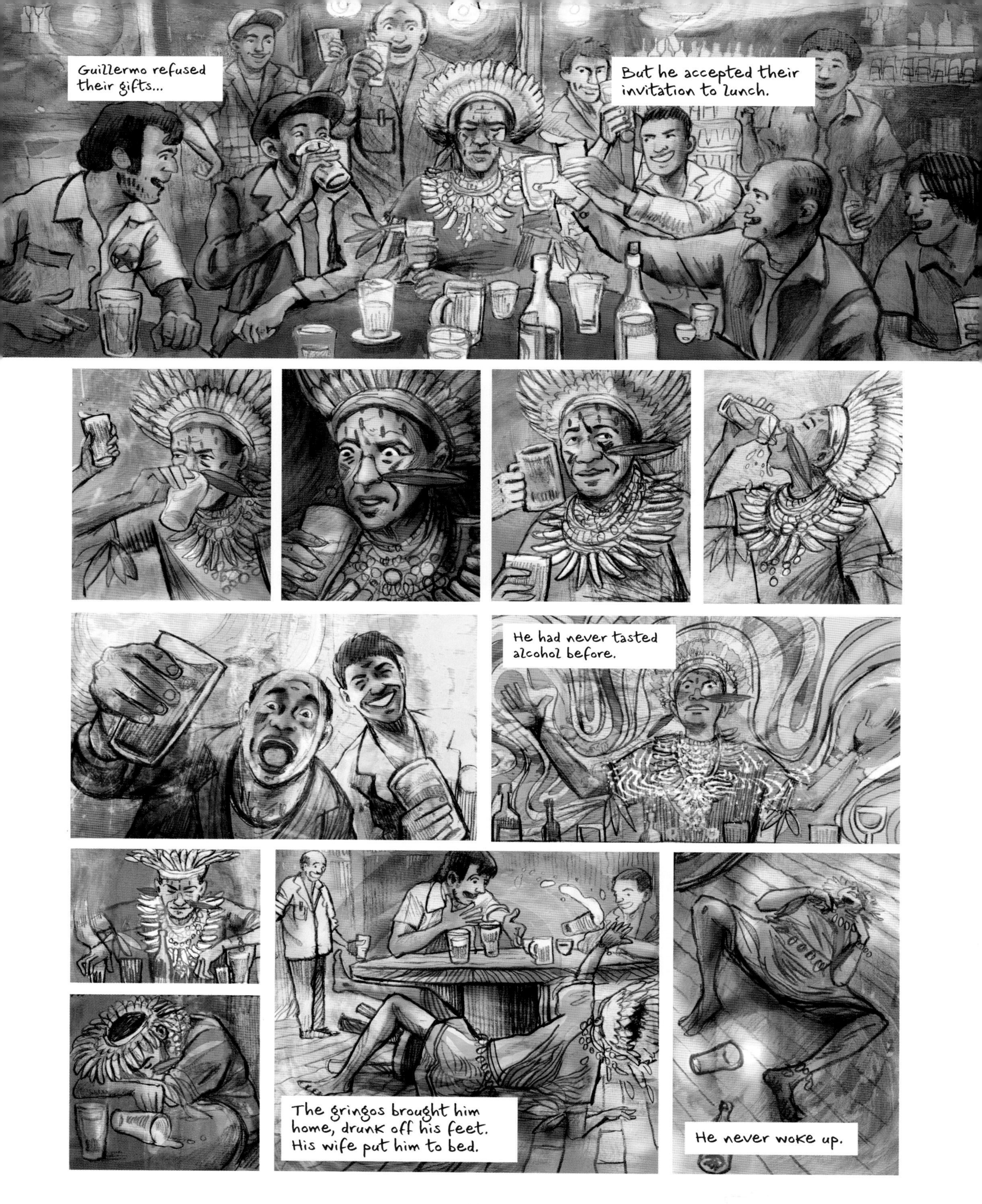
Guillermo refused their gifts...
But he accepted their invitation to lunch.
He had never tasted alcohol before.
The gringos brought him home, drunk off his feet. His wife put him to bed.
He never woke up.

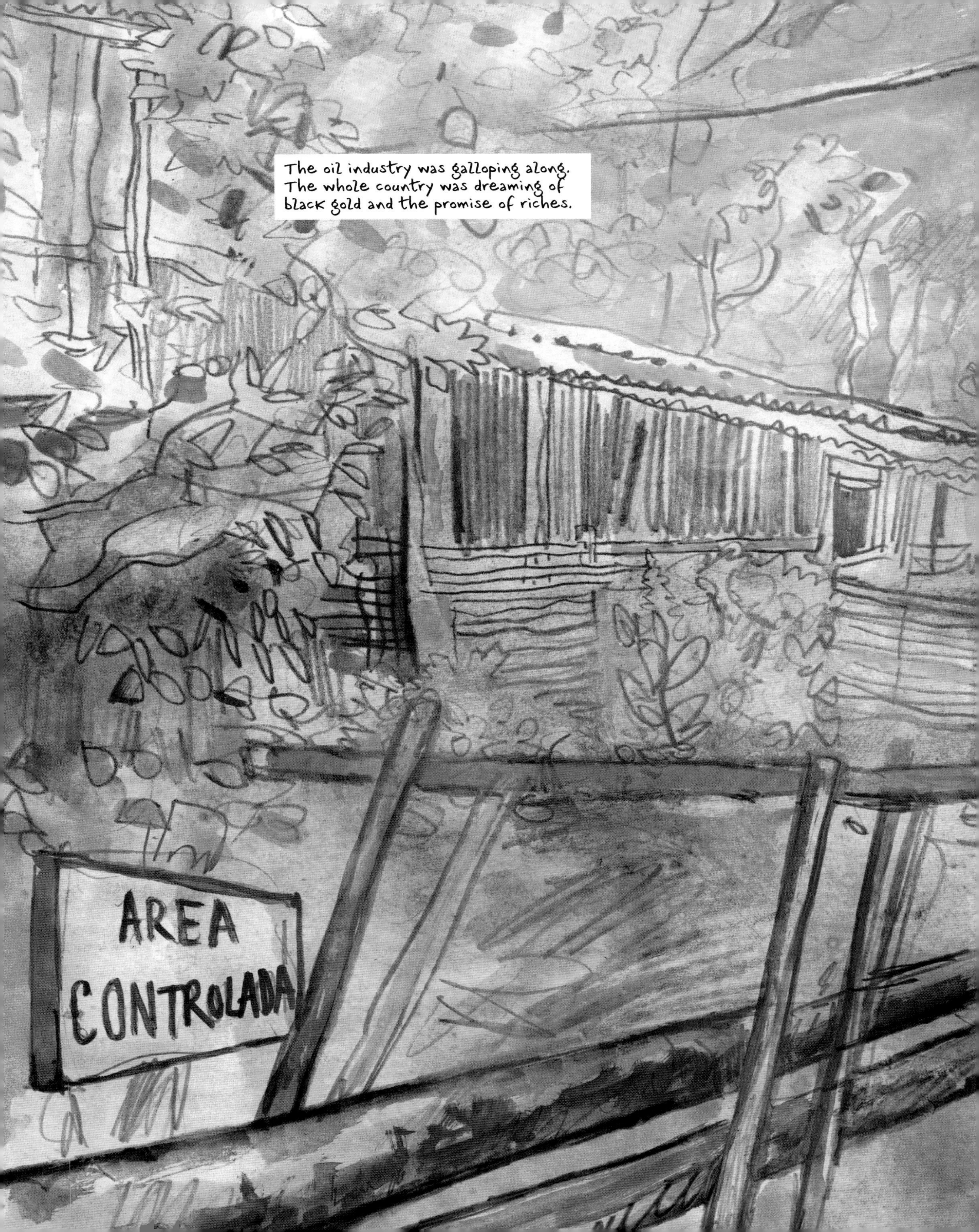
The oil industry was galloping along. The whole country was dreaming of black gold and the promise of riches.
AREA CONTROLADA

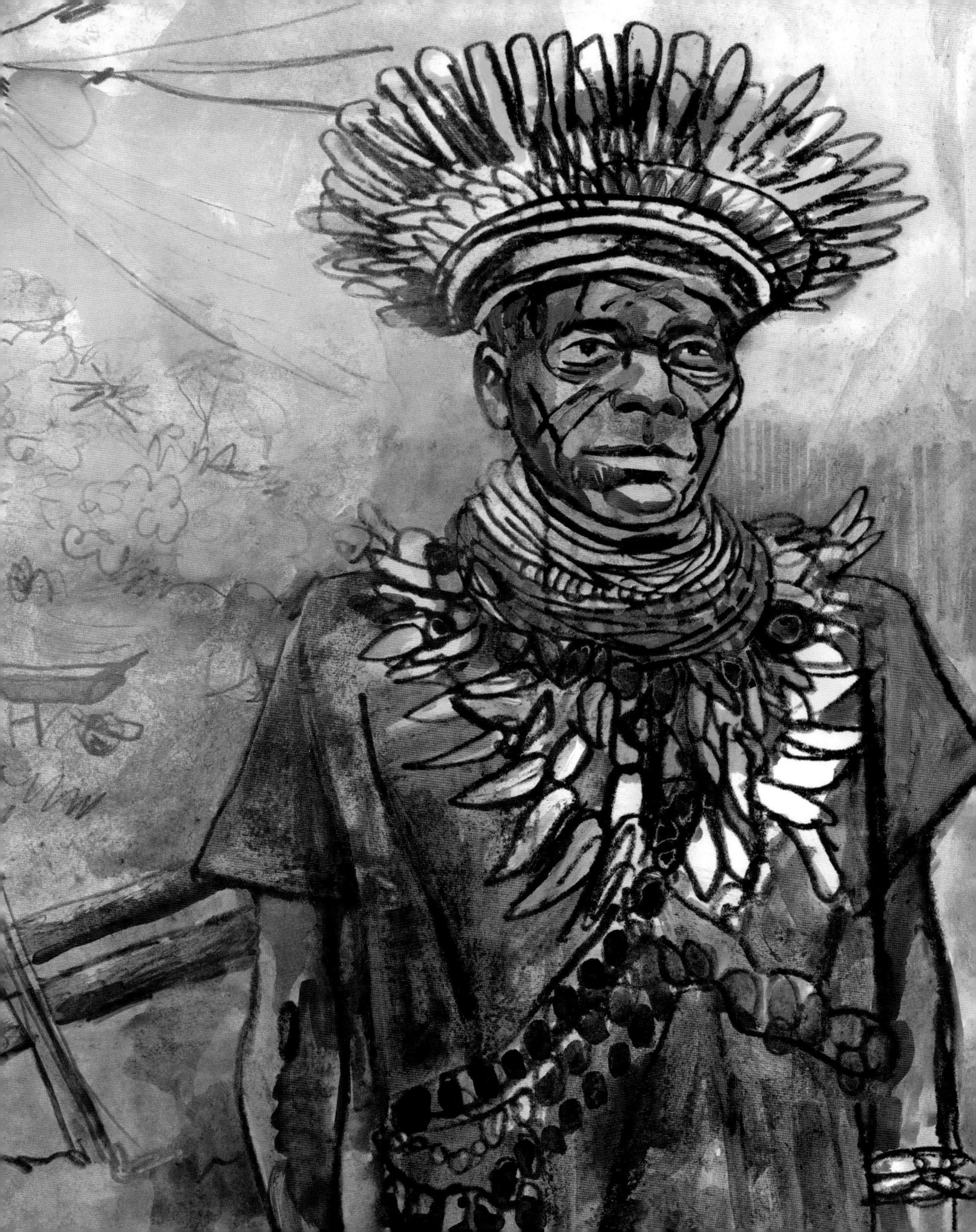

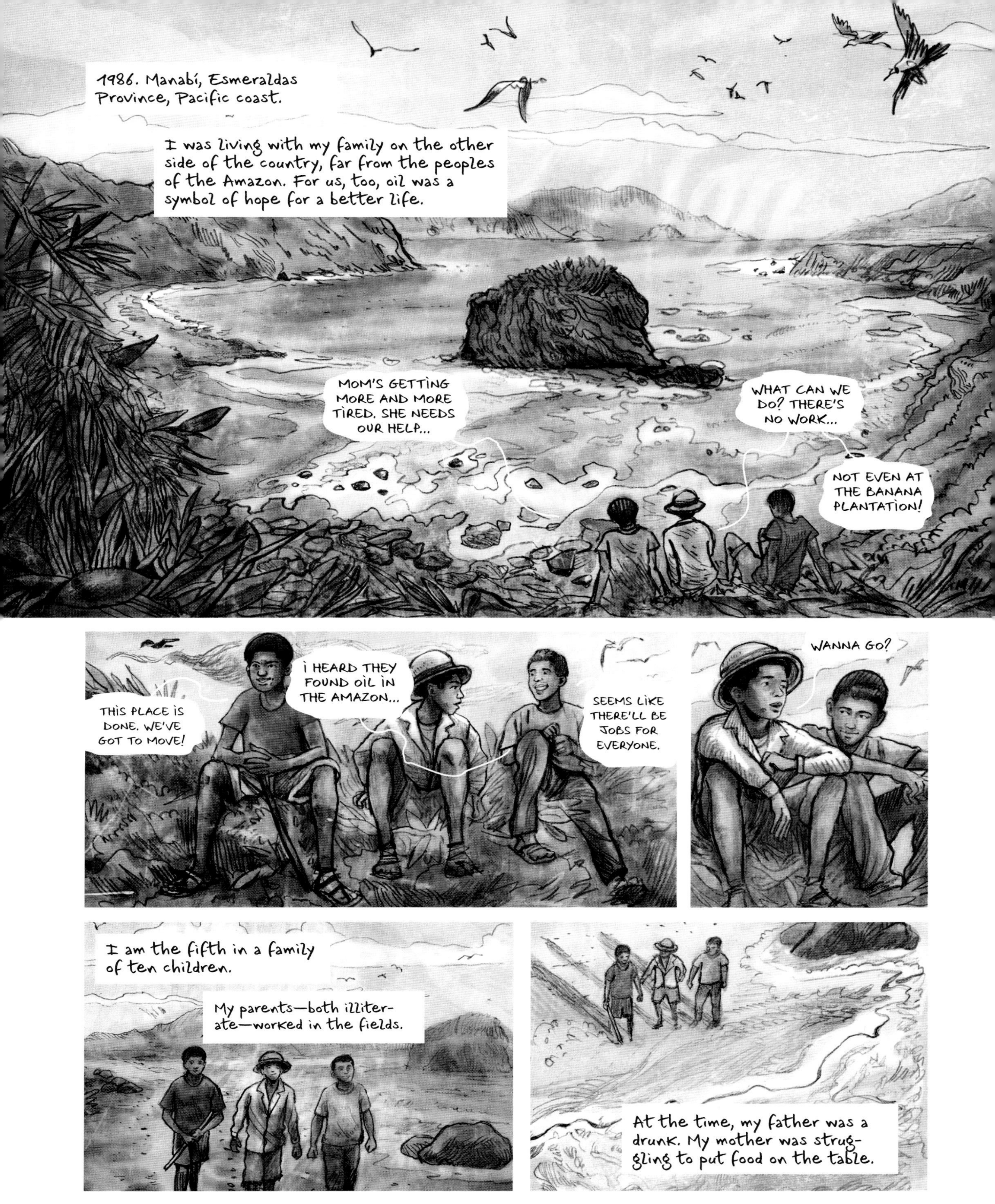
1986. Manabí, Esmeraldas Province, Pacific coast.
I was living with my family on the other side of the country, far from the peoples of the Amazon. For us, too, oil was a symbol of hope for a better life.
MOM'S GETTING MORE AND MORE TIRED. SHE NEEDS OUR HELP...
WHAT CAN WE DO? THERE'S NO WORK...
NOT EVEN AT THE BANANA PLANTATION!
THIS PLACE IS DONE. WE'VE GOT TO MOVE!
I HEARD THEY FOUND OIL IN THE AMAZON...
SEEMS LIKE THERE'LL BE JOBS FOR EVERYONE.
WANNA GO?
I am the fifth in a family of ten children.
My parents—both illiterate—worked in the fields.
At the time, my father was a drunk. My mother was struggling to put food on the table.

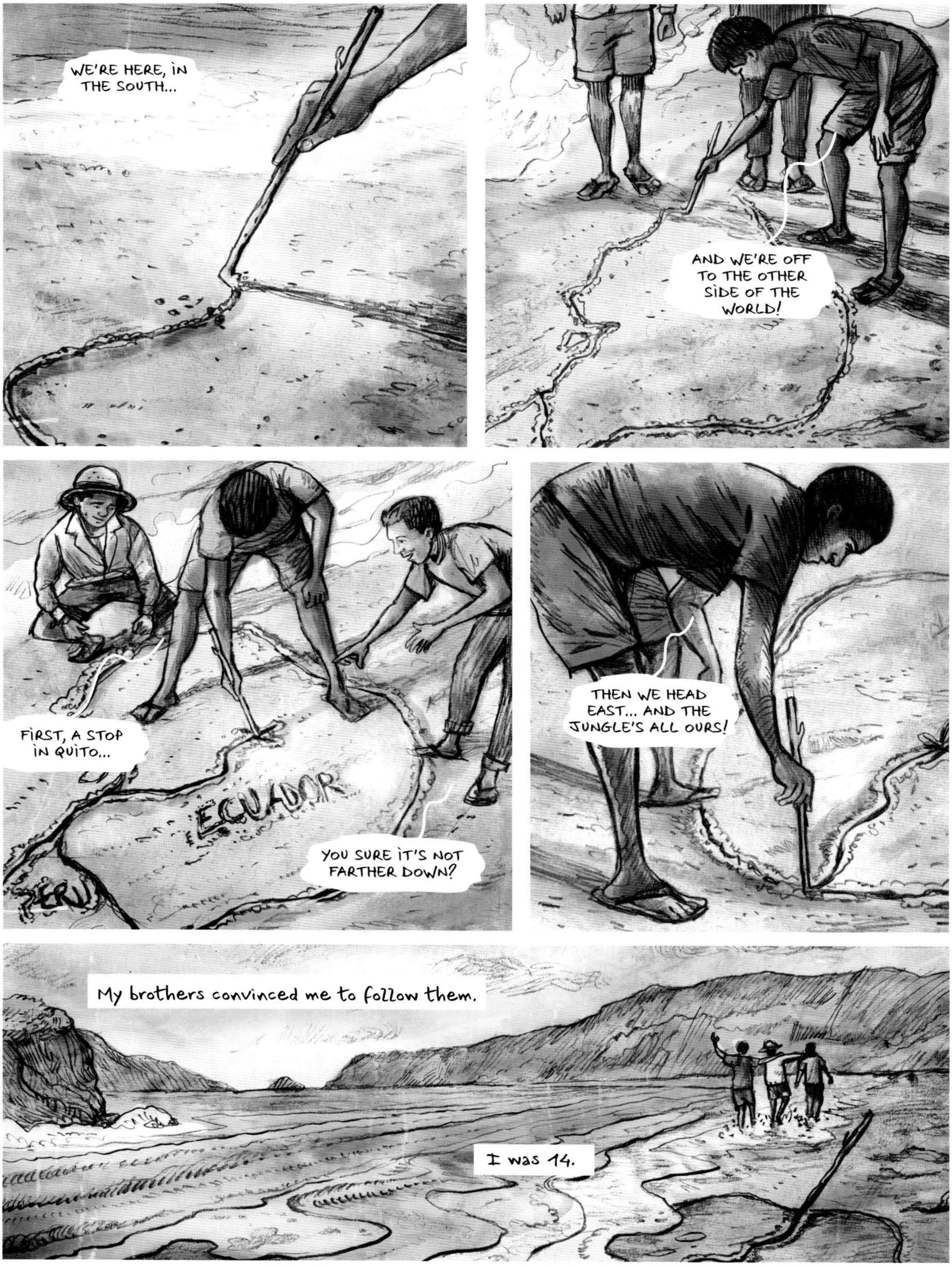
WE'RE HERE, IN THE SOUTH...
AND WE'RE OFF TO THE OTHER SIDE OF THE WORLD!
FIRST, A STOP IN QUITO...
ECUADOR
PERU
YOU SURE IT'S NOT FARTHER DOWN?
THEN WE HEAD EAST... AND THE JUNGLE'S ALL OURS!
My brothers convinced me to follow them.
I was 14.

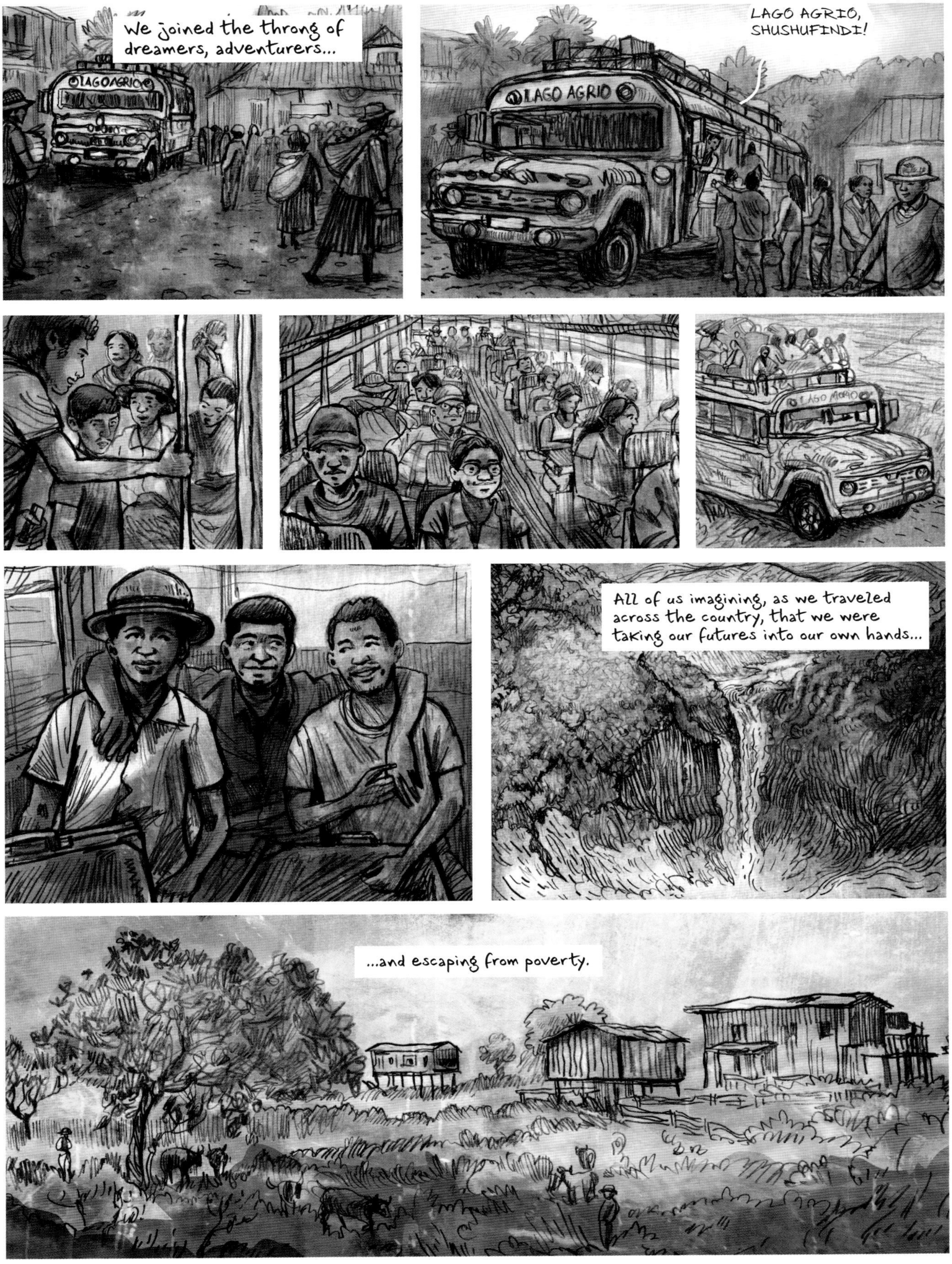
We joined the throng of dreamers, adventurers...
LAGO AGRIO
LAGO AGRIO
LAGO AGRIO, SHUSHUFINDI!
LAGO AGRIO
All of us imagining, as we traveled across the country, that we were taking our futures into our own hands...
...and escaping from poverty.

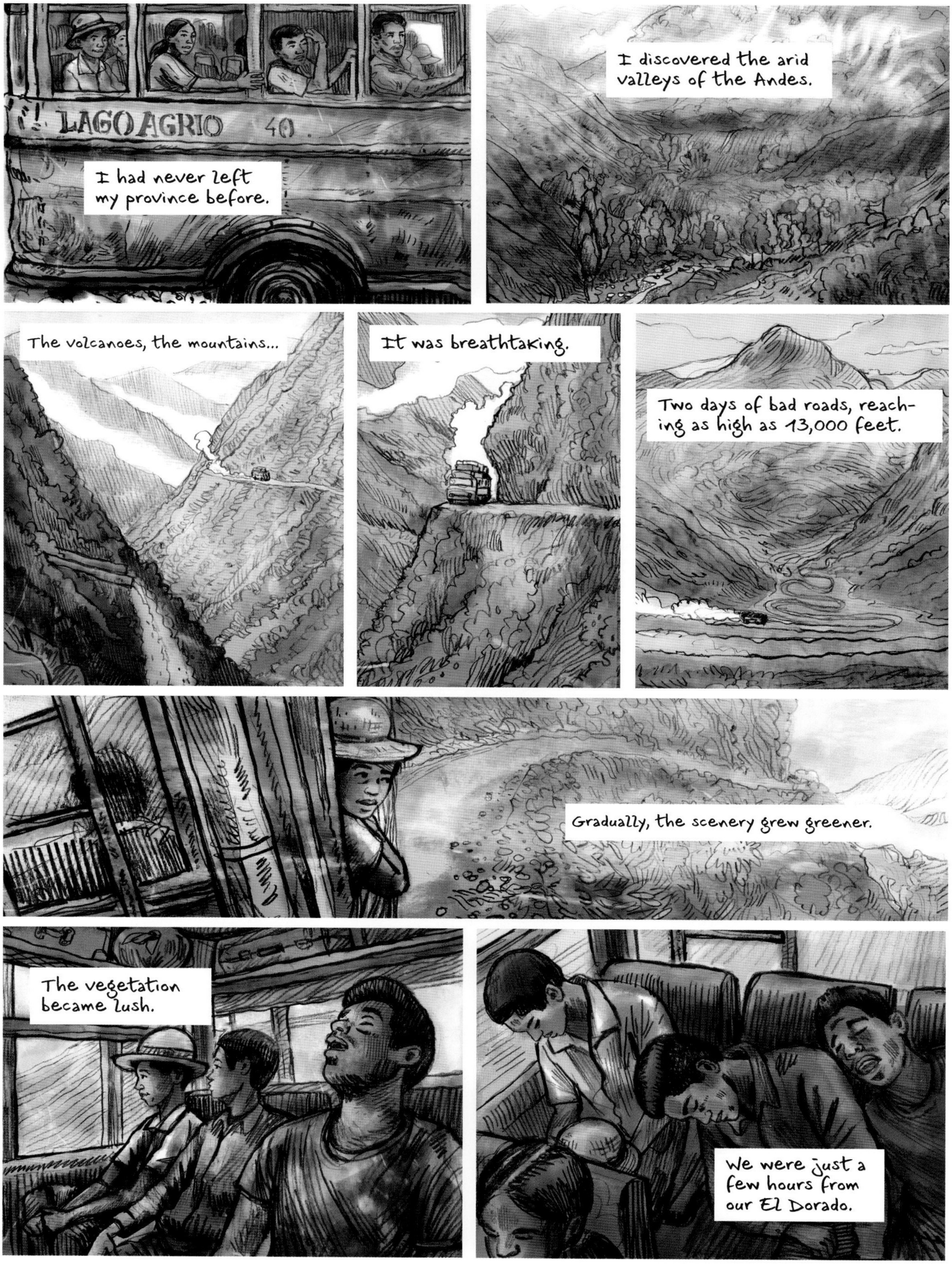
LAGO AGRIO 40
I had never left my province before.
I discovered the arid valleys of the Andes.
The volcanoes, the mountains...
It was breathtaking.
Two days of bad roads, reaching as high as 13,000 feet.
Gradually, the scenery grew greener.
The vegetation became lush.
We were just a few hours from our El Dorado.

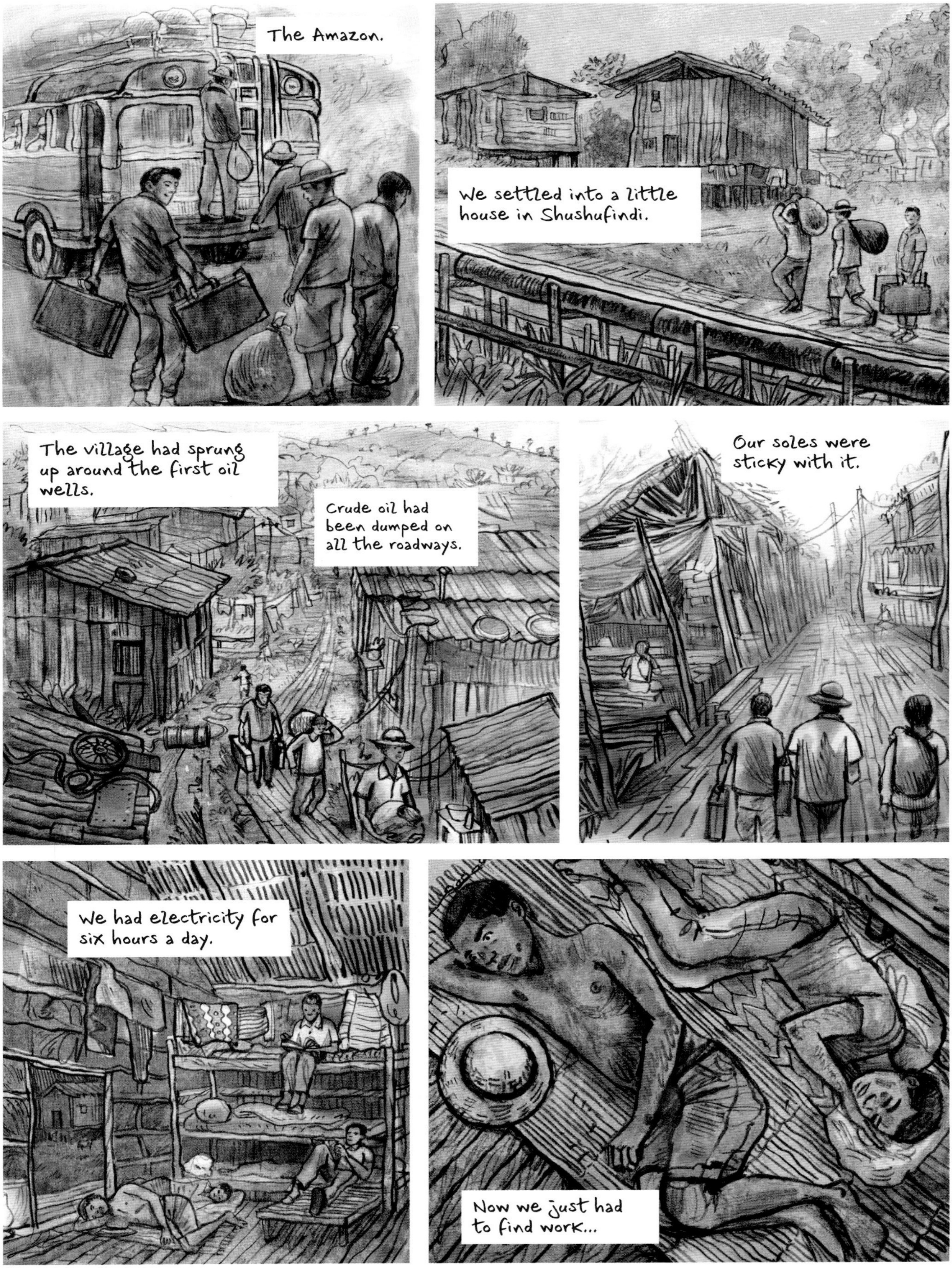
The Amazon.
We settled into a little house in Shushufindi.
The village had sprung up around the first oil wells.
Crude oil had been dumped on all the roadways.
Our soles were sticky with it.
We had electricity for six hours a day.
Now we just had to find work...

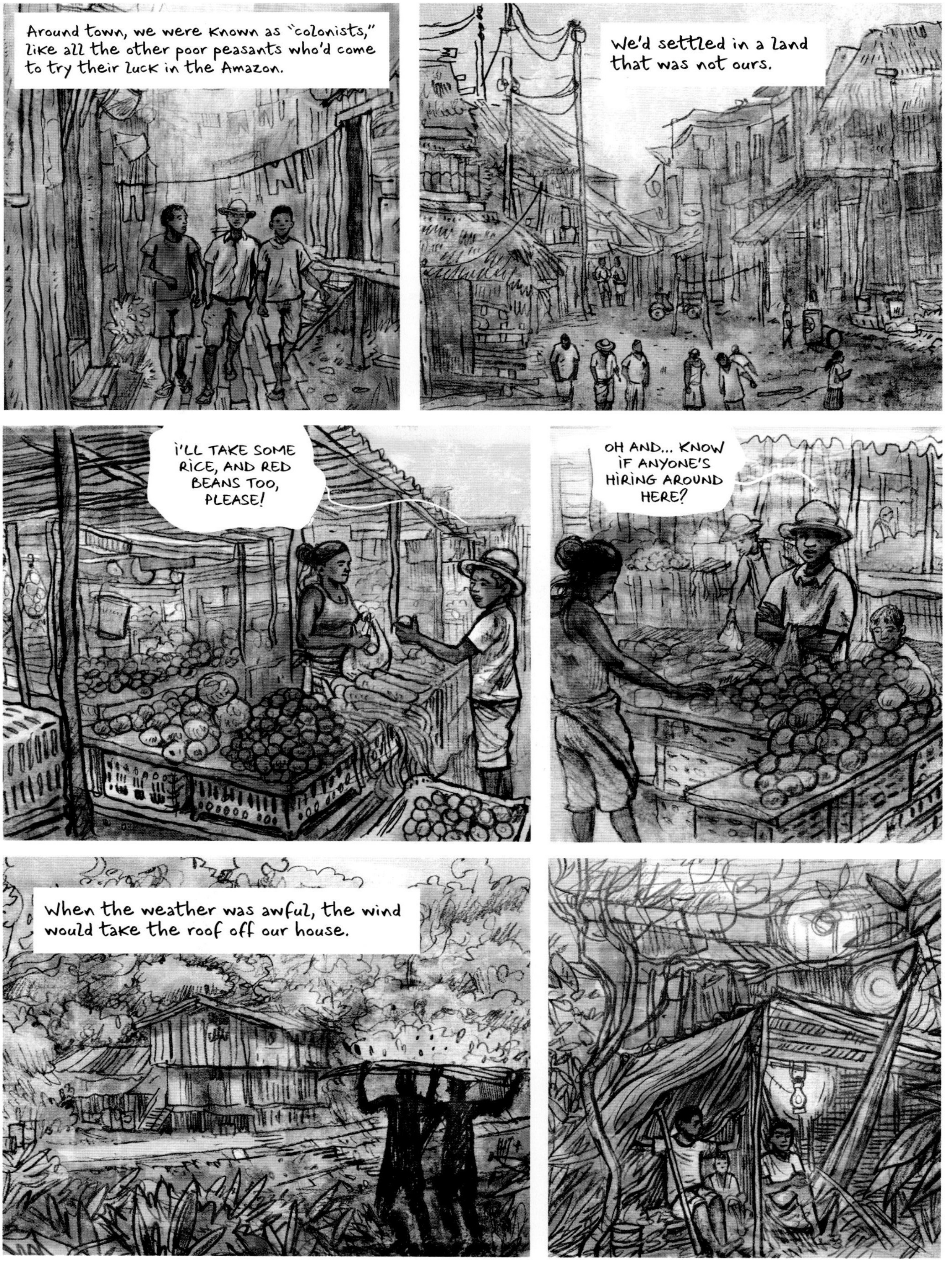
Around town, we were known as "colonists," like all the other poor peasants who'd come to try their luck in the Amazon.
We'd settled in a land that was not ours.
I'LL TAKE SOME RICE, AND RED BEANS TOO, PLEASE!
OH AND... KNOW IF ANYONE'S HIRING AROUND HERE?
When the weather was awful, the wind would take the roof off our house.

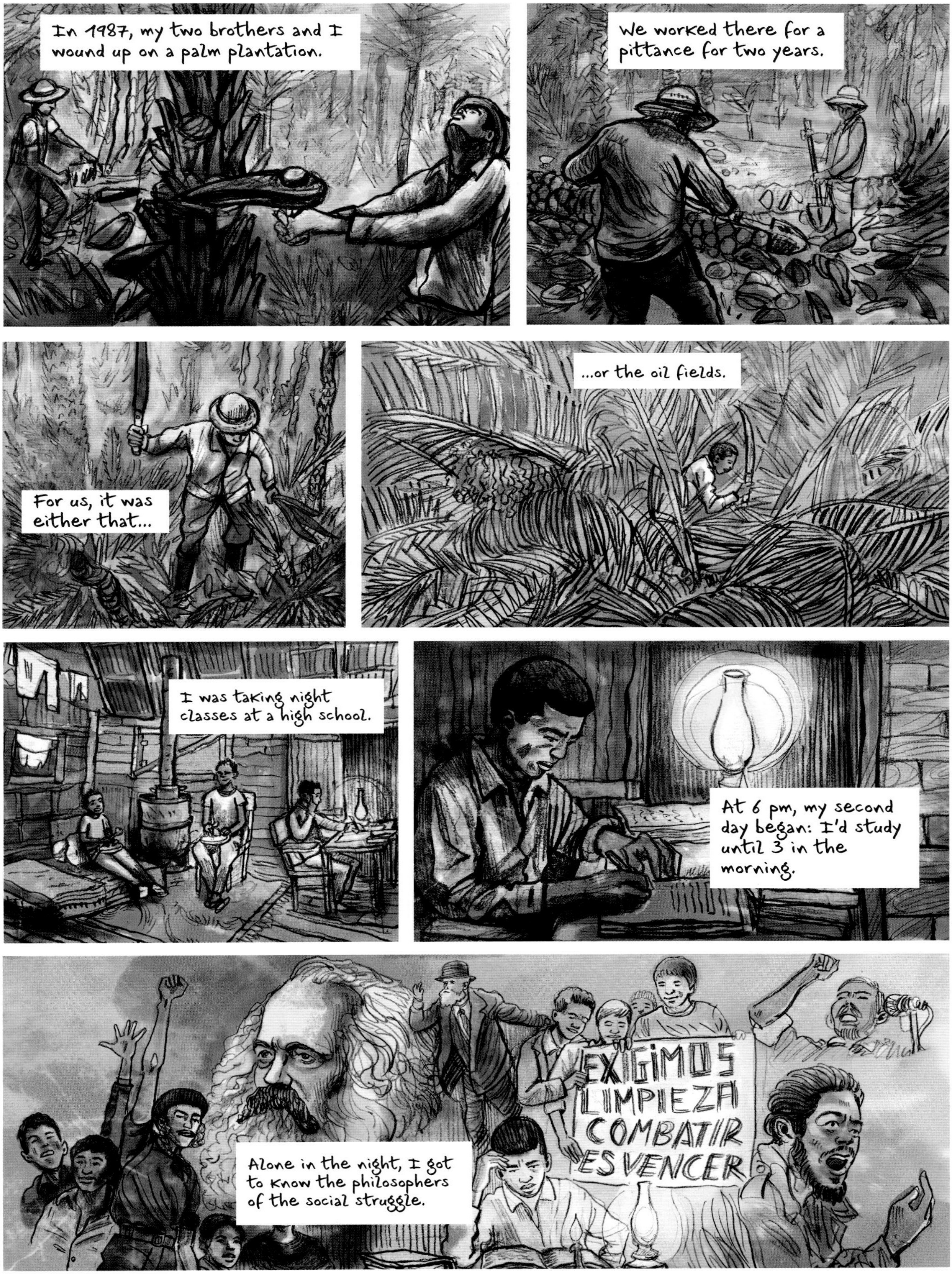
In 1987, my two brothers and I wound up on a palm plantation.
We worked there for a pittance for two years.
For us, it was either that...
...or the oil fields.
I was taking night classes at a high school.
At 6 pm, my second day began: I'd study until 3 in the morning.
Alone in the night, I got to know the philosophers of the social struggle.
EXIGIMOS LIMPIEZA COMBATIR ES VENCER

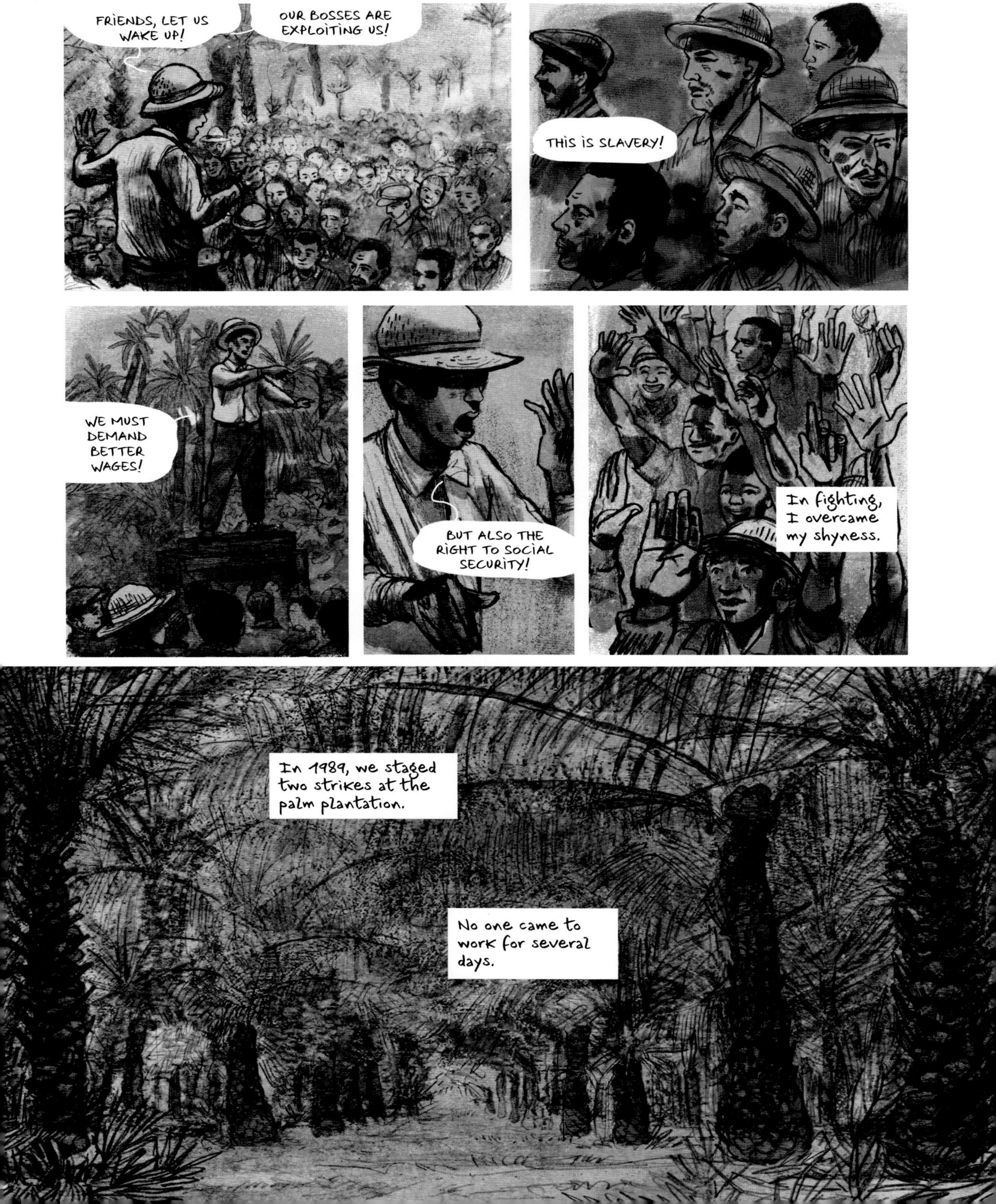
FRIENDS, LET US WAKE UP!
OUR BOSSES ARE EXPLOITING US!
THIS IS SLAVERY!
WE MUST DEMAND BETTER WAGES!
BUT ALSO THE RIGHT TO SOCIAL SECURITY!
In fighting, I overcame my shyness.
In 1989, we staged two strikes at the palm plantation.
No one came to work for several days.

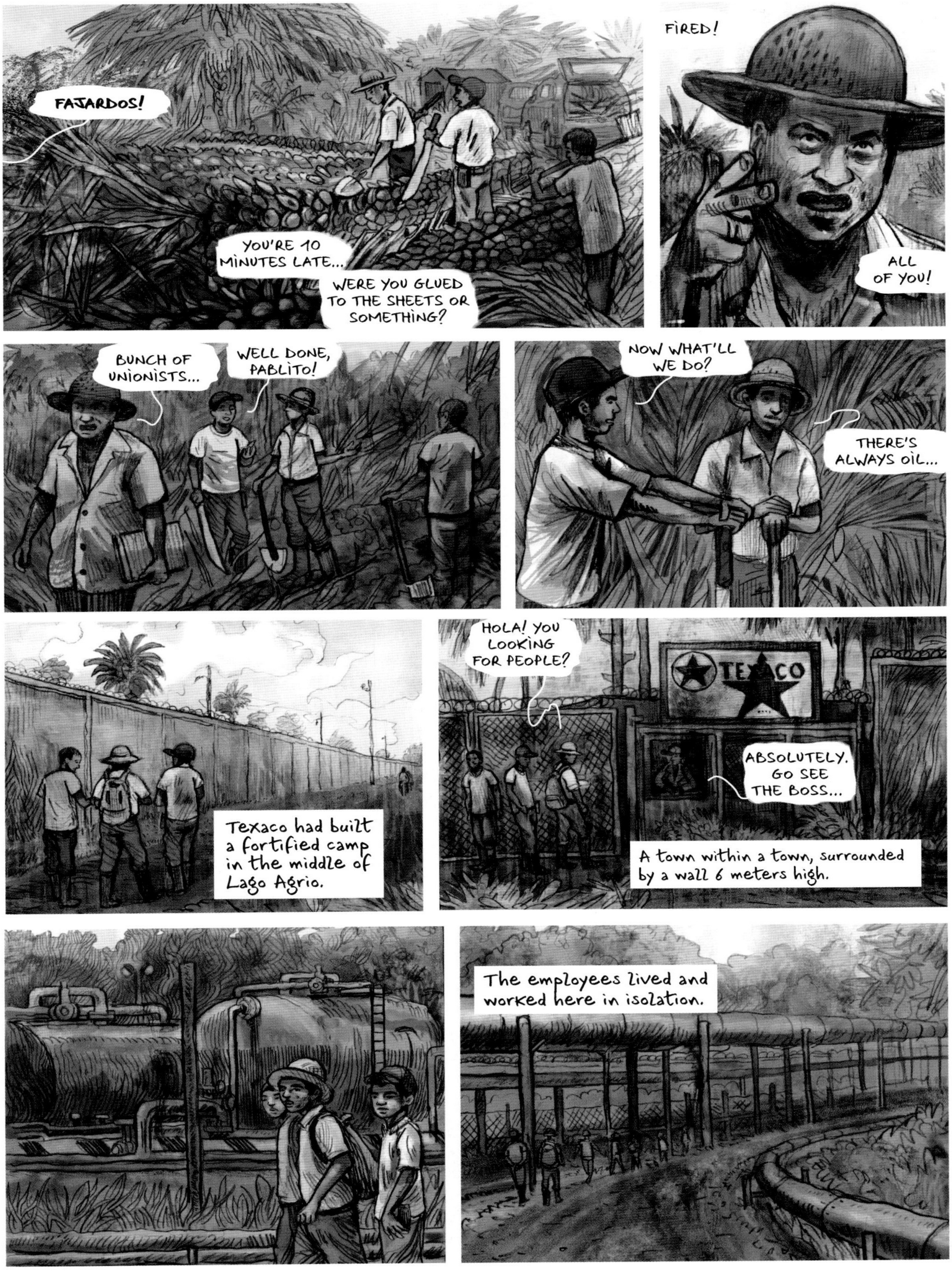
FAJARDOS!
YOU'RE 10 MINUTES LATE...
WERE YOU GLUED TO THE SHEETS OR SOMETHING?
FIRED!
ALL OF YOU!
BUNCH OF UNIONISTS...
WELL DONE, PABLITO!
NOW WHAT'LL WE DO?
THERE'S ALWAYS OIL...
Texaco had built a fortified camp in the middle of Lago Agrio.
HOLA! YOU LOOKING FOR PEOPLE?
TEXACO
ABSOLUTELY. GO SEE THE BOSS...
A town within a town, surrounded by a wall 6 meters high.
The employees lived and worked here in isolation.

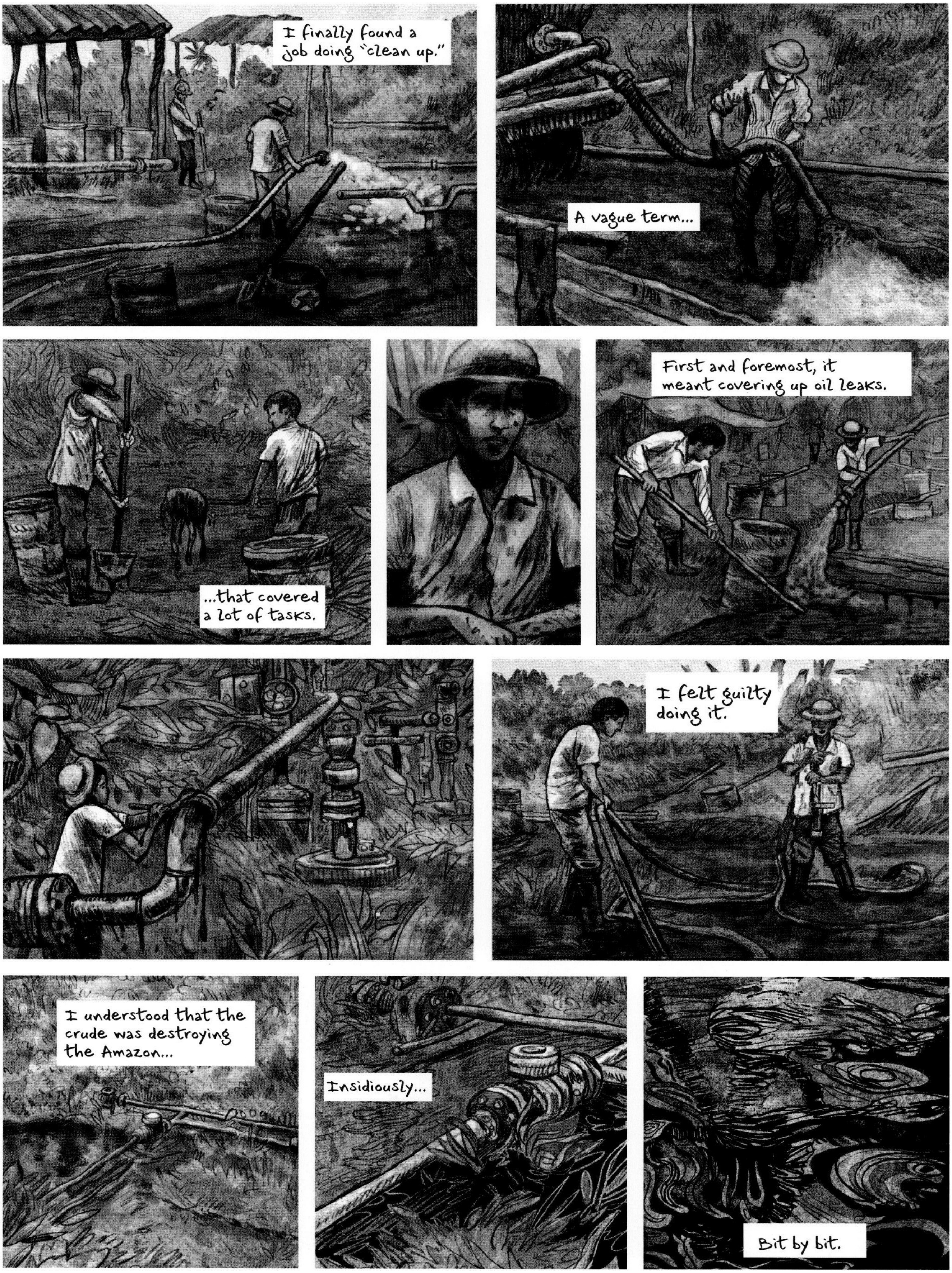
I finally found a job doing "clean up."
A vague term...
...that covered a lot of tasks.
First and foremost, it meant covering up oil leaks.
I felt guilty doing it.
I understood that the crude was destroying the Amazon...
Insidiously...
Bit by bit.

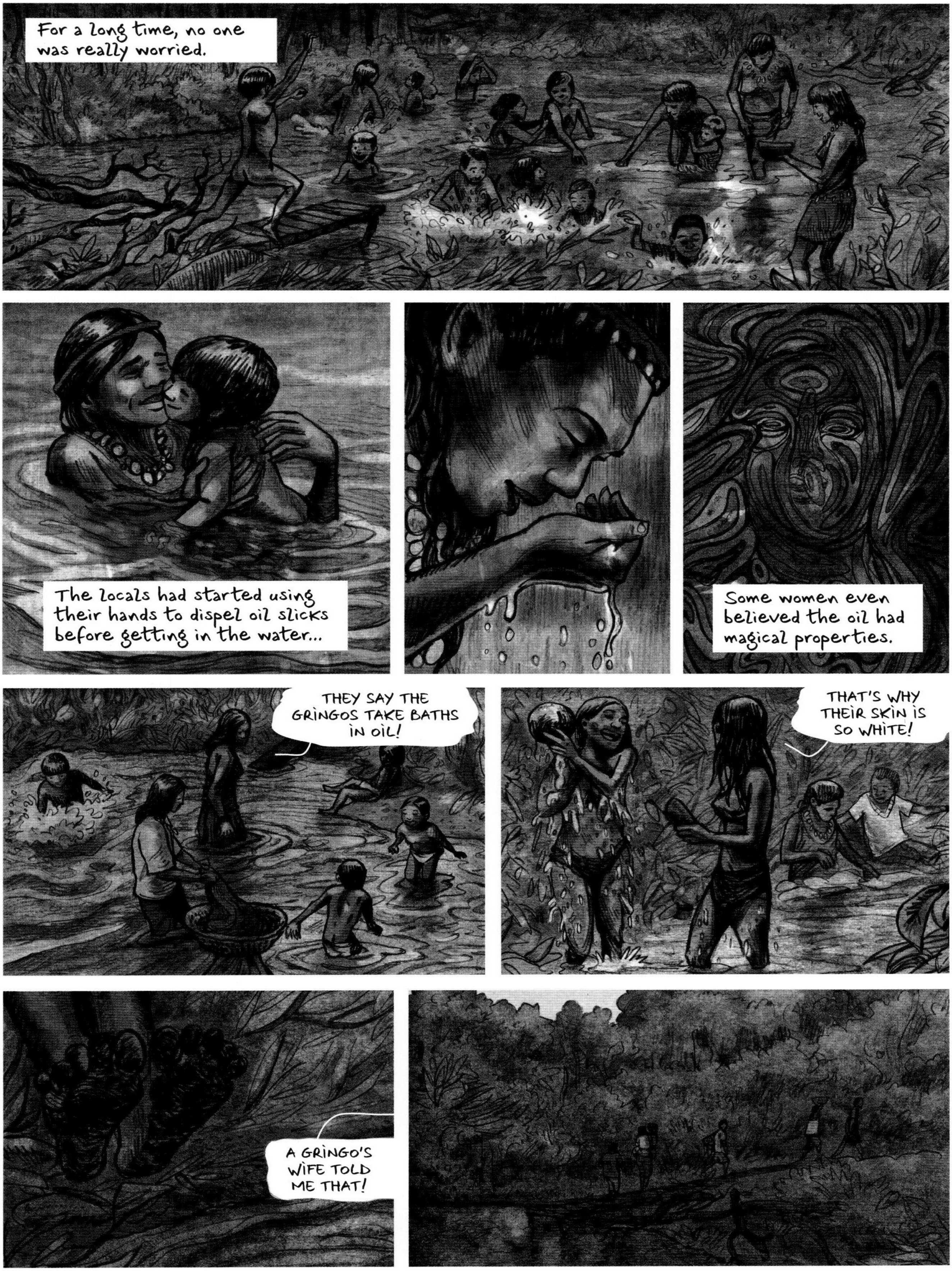
For a long time, no one was really worried.
The locals had started using their hands to dispel oil slicks before getting in the water...
Some women even believed the oil had magical properties.
THEY SAY THE GRINGOS TAKE BATHS IN OIL!
THAT'S WHY THEIR SKIN IS SO WHITE!
A GRINGO'S WIFE TOLD ME THAT!

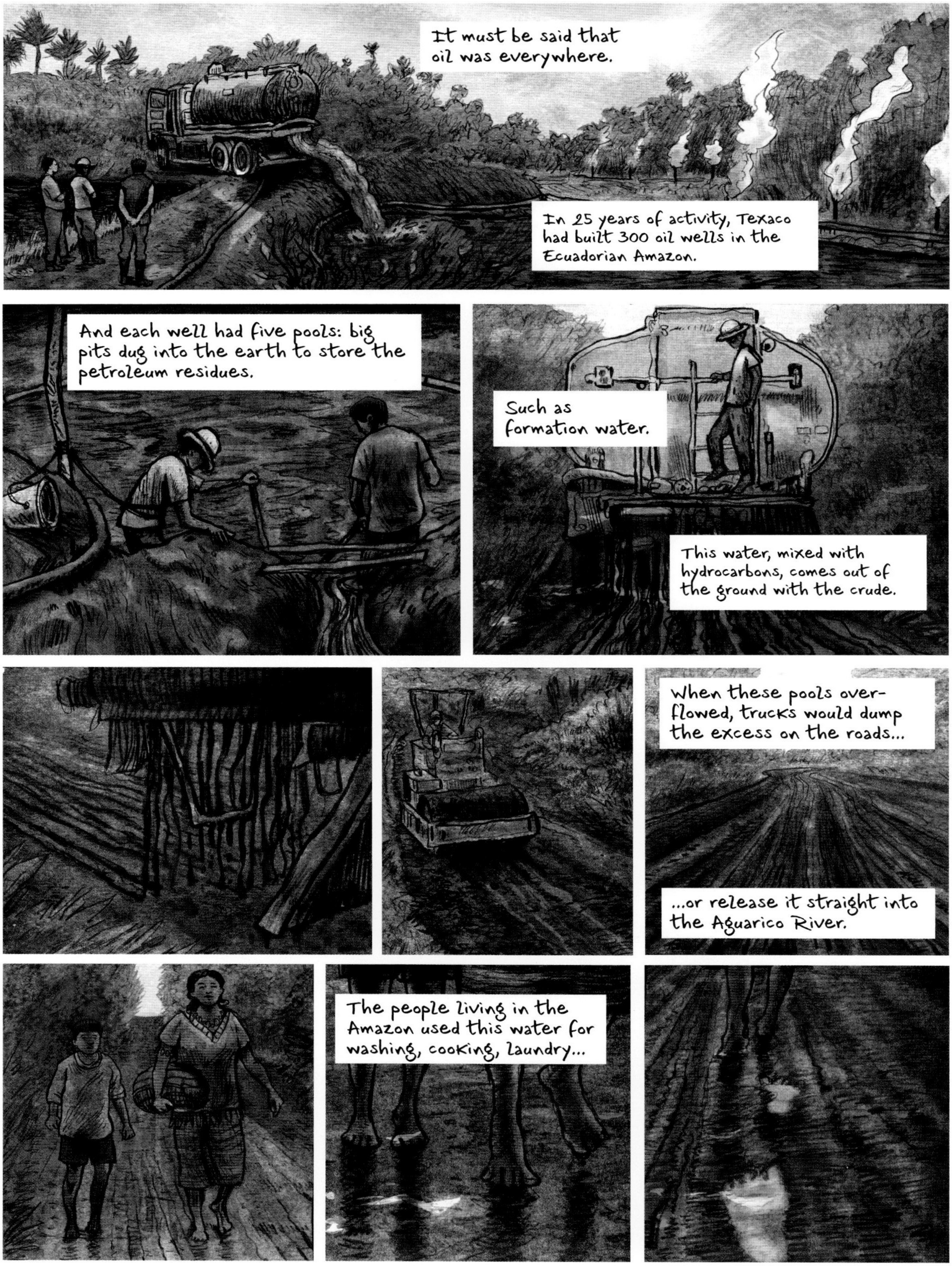
It must be said that oil was everywhere.
In 25 years of activity, Texaco had built 300 oil wells in the Ecuadorian Amazon.
And each well had five pools: big pits dug into the earth to store the petroleum residues.
Such as formation water.
This water, mixed with hydrocarbons, comes out of the ground with the crude.
When these pools over-flowed, trucks would dump the excess on the roads...
...or release it straight into the Aguarico River.
The people living in the Amazon used this water for washing, cooking, laundry...

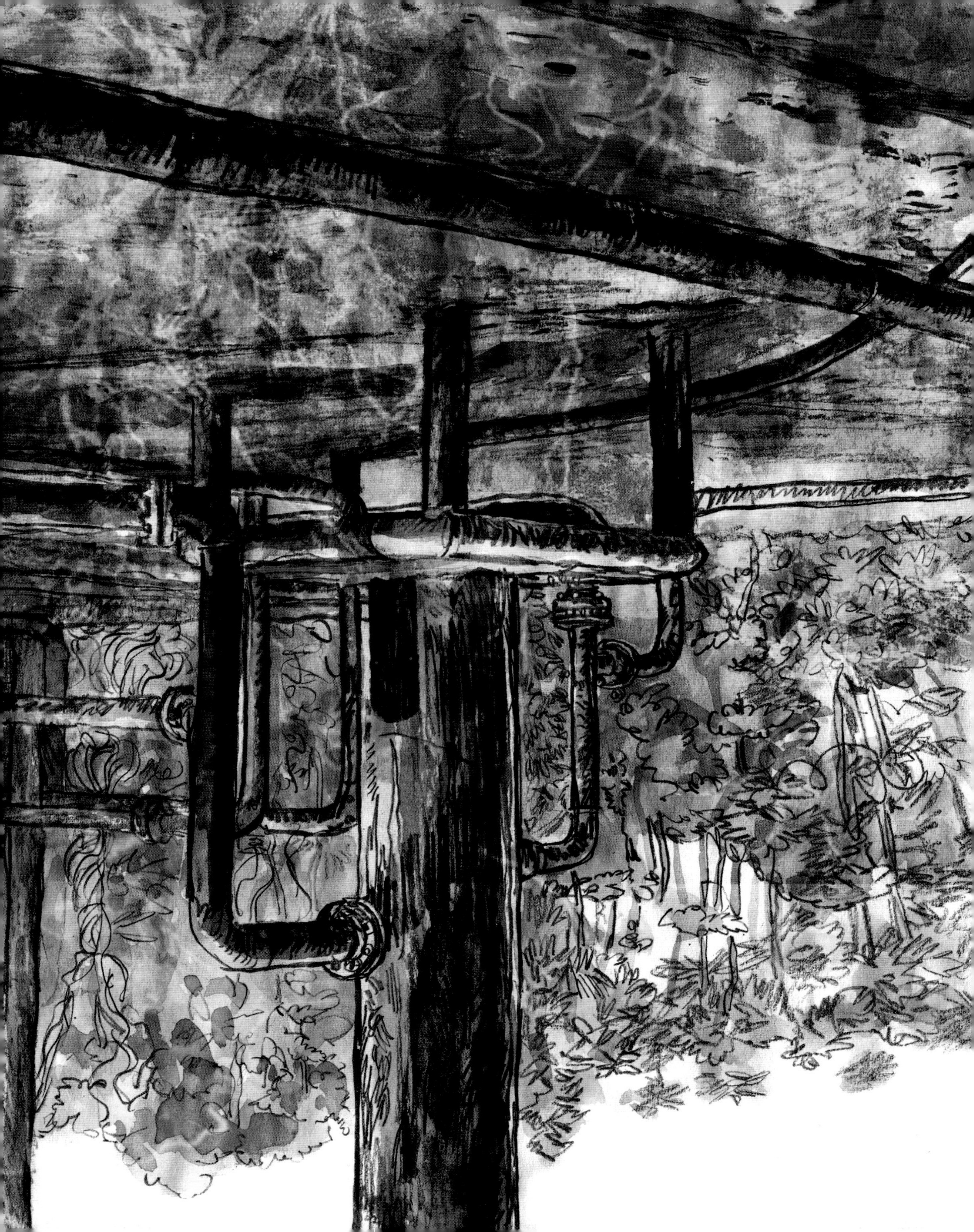

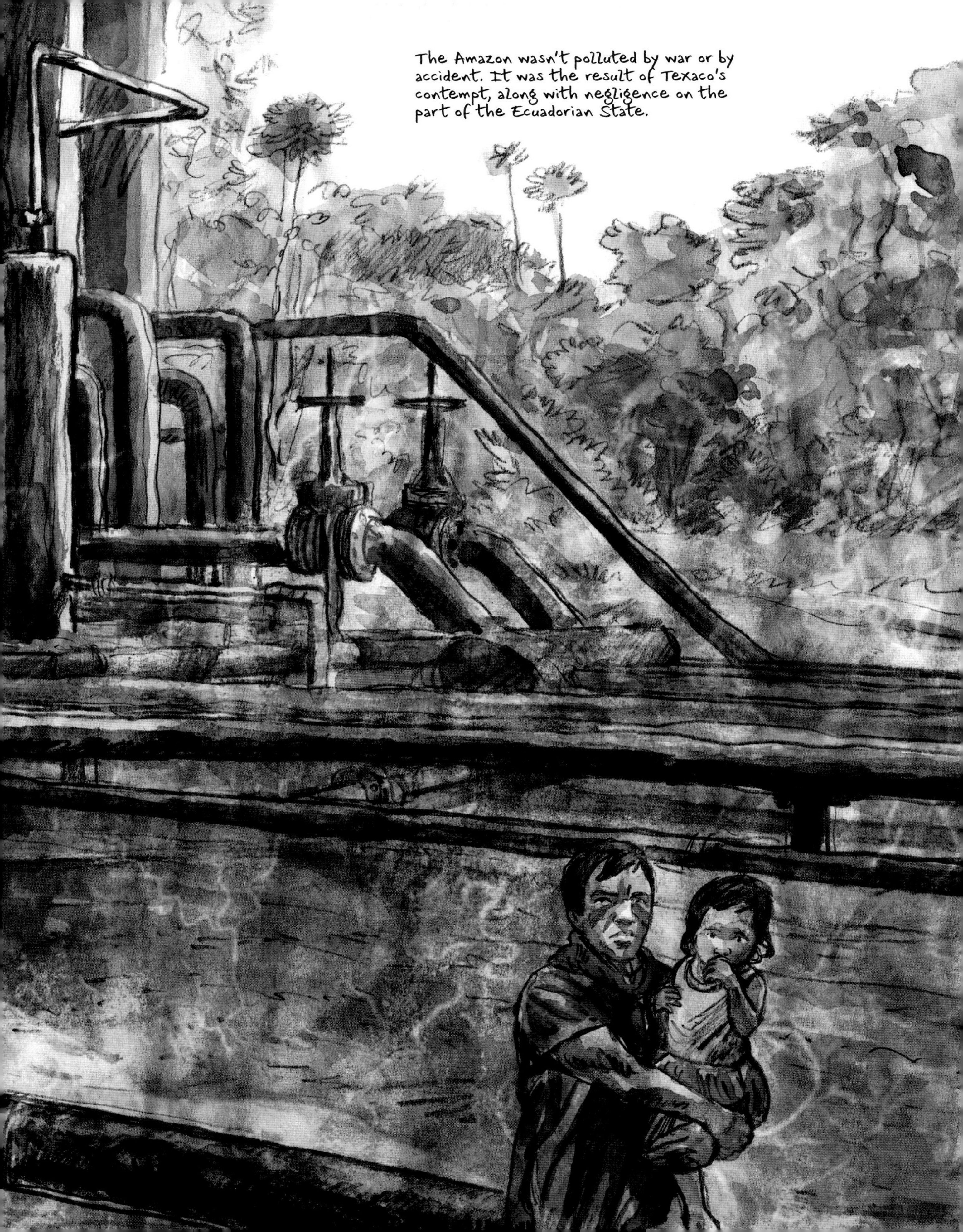
The Amazon wasn't polluted by war or by accident. It was the result of Texaco's contempt, along with negligence on the part of the Ecuadorian State.

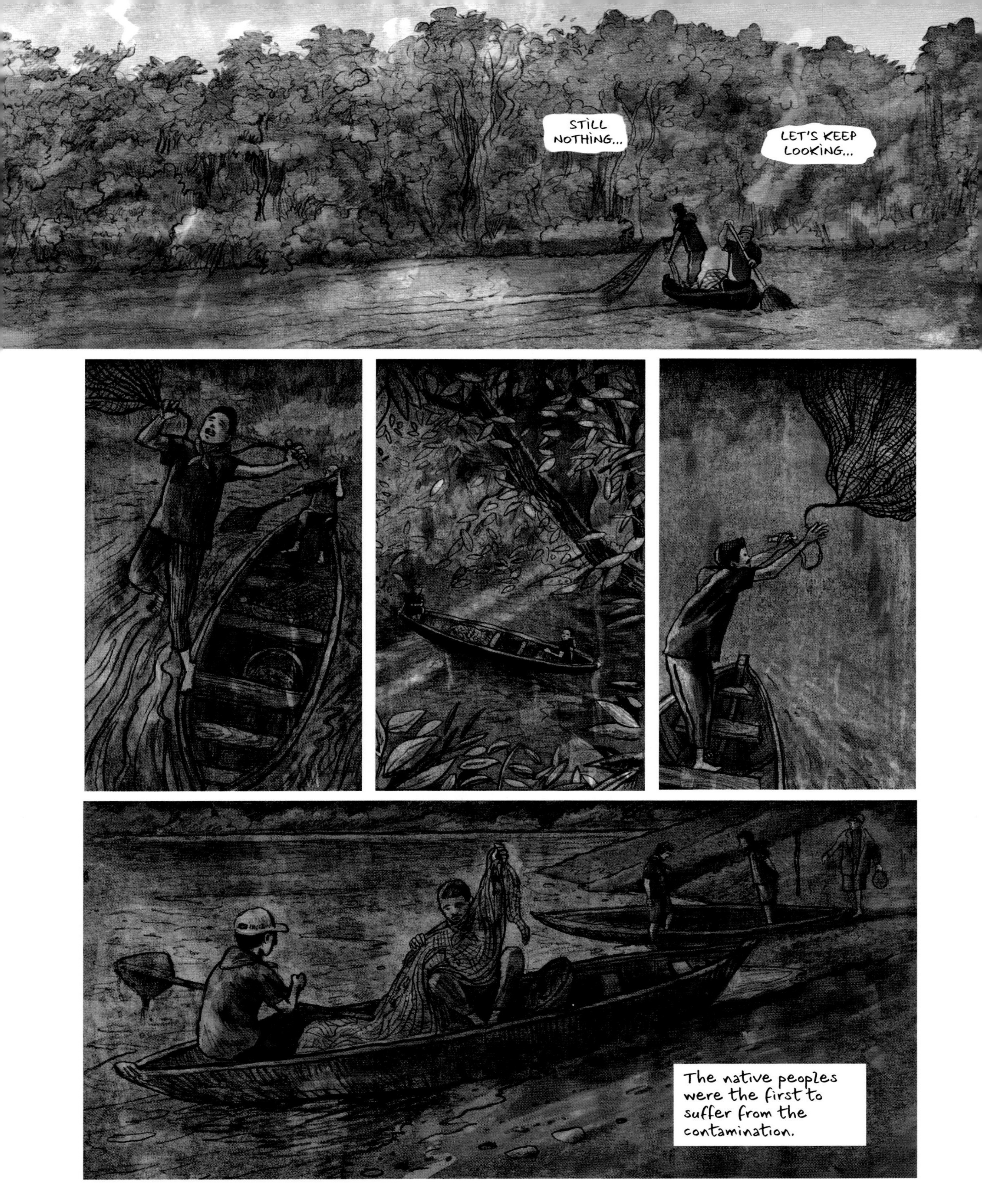
STILL NOTHING...
LET'S KEEP LOOKING...
The native peoples were the first to suffer from the contamination.

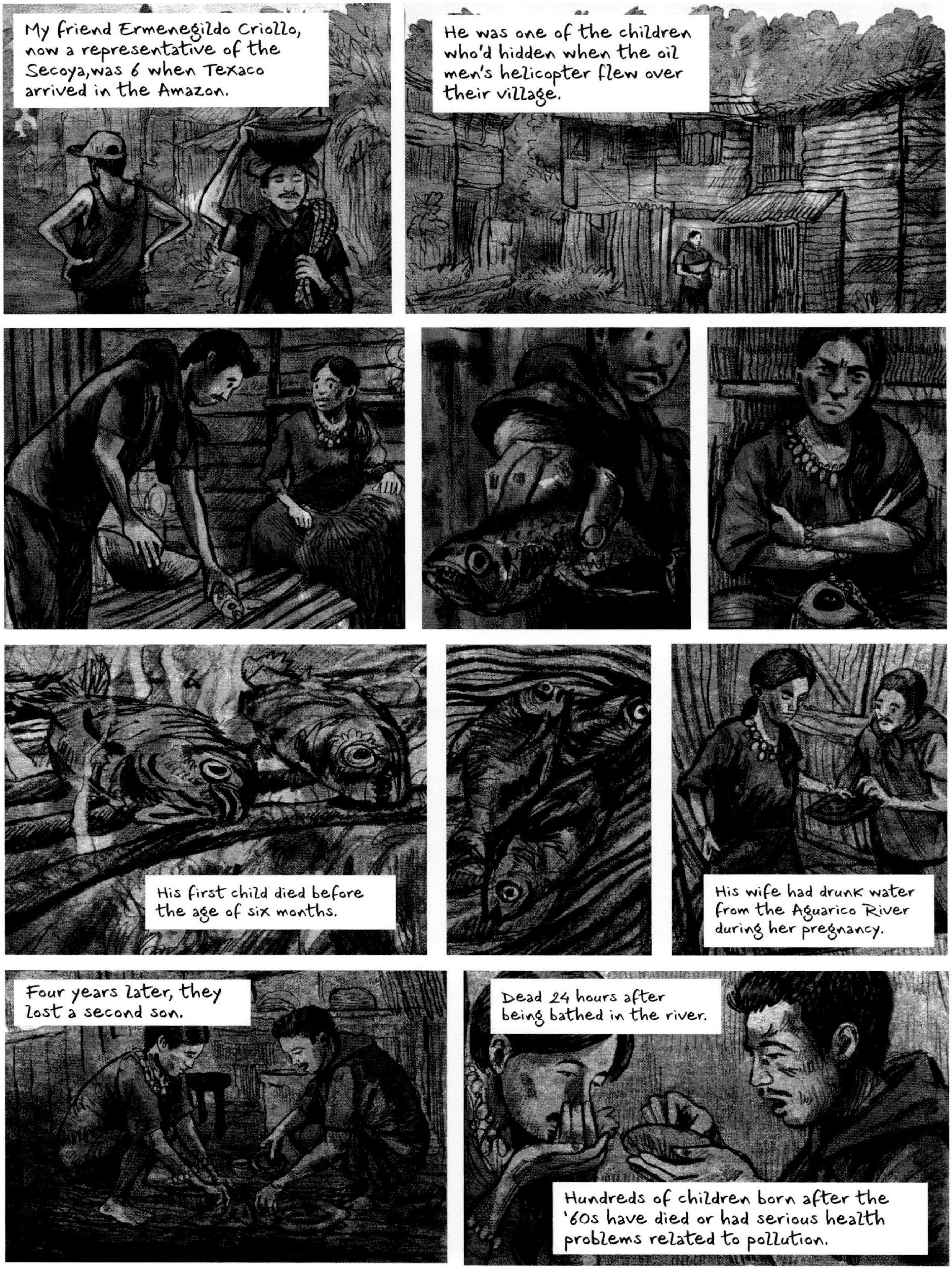
My friend Ermenegildo Criollo, now a representative of the Secoya, was 6 when Texaco arrived in the Amazon.
He was one of the children who'd hidden when the oil men's helicopter flew over their village.
His first child died before the age of six months.
His wife had drunk water from the Aguarico River during her pregnancy.
Four years later, they lost a second son.
Dead 24 hours after being bathed in the river.
Hundreds of children born after the '60s have died or had serious health problems related to pollution.

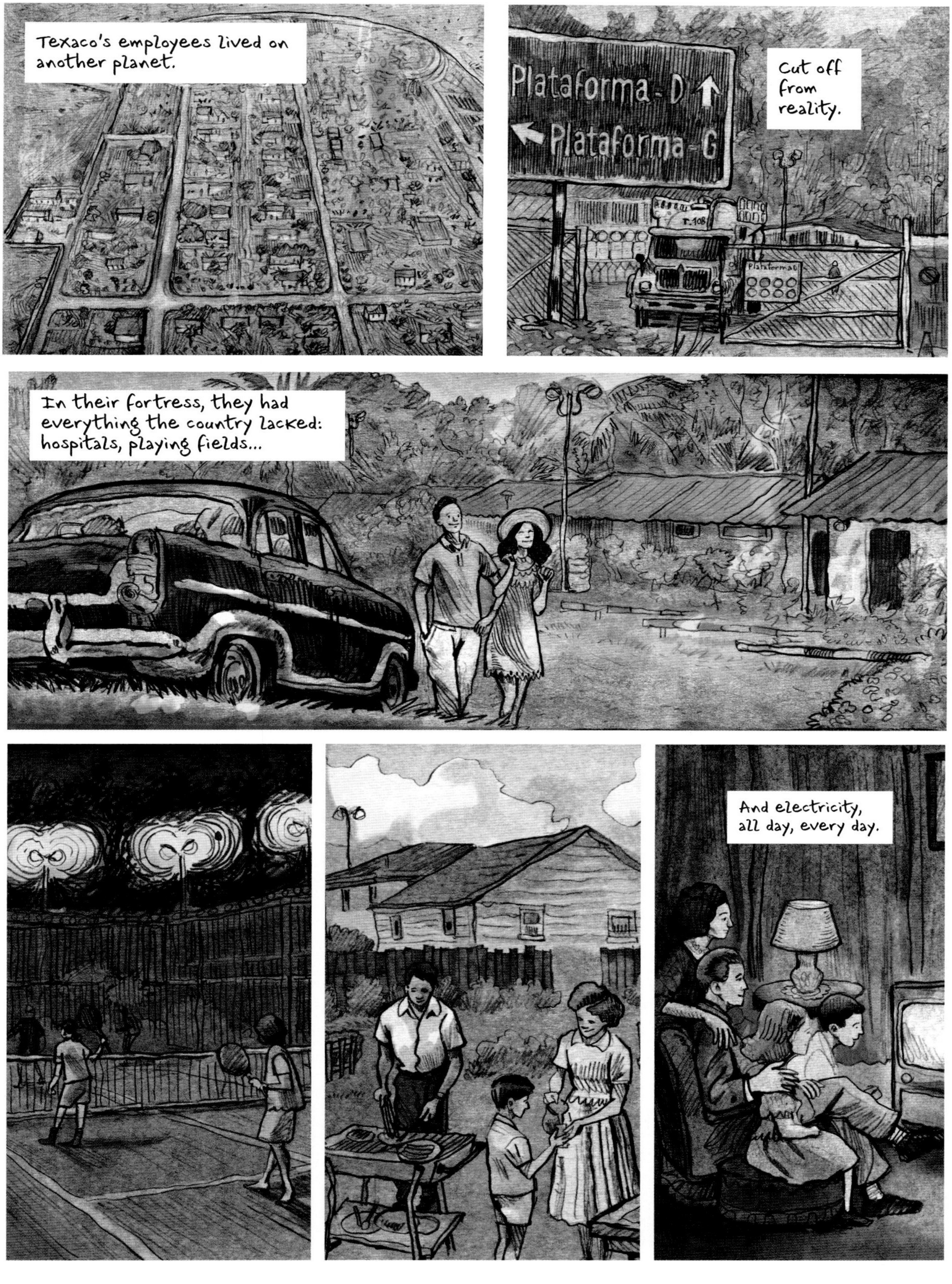
Texaco's employees lived on another planet.
Plataforma-D
Plataforma-G
Cut off from reality.
In their fortress, they had everything the country lacked: hospitals, playing fields...
And electricity, all day, every day.

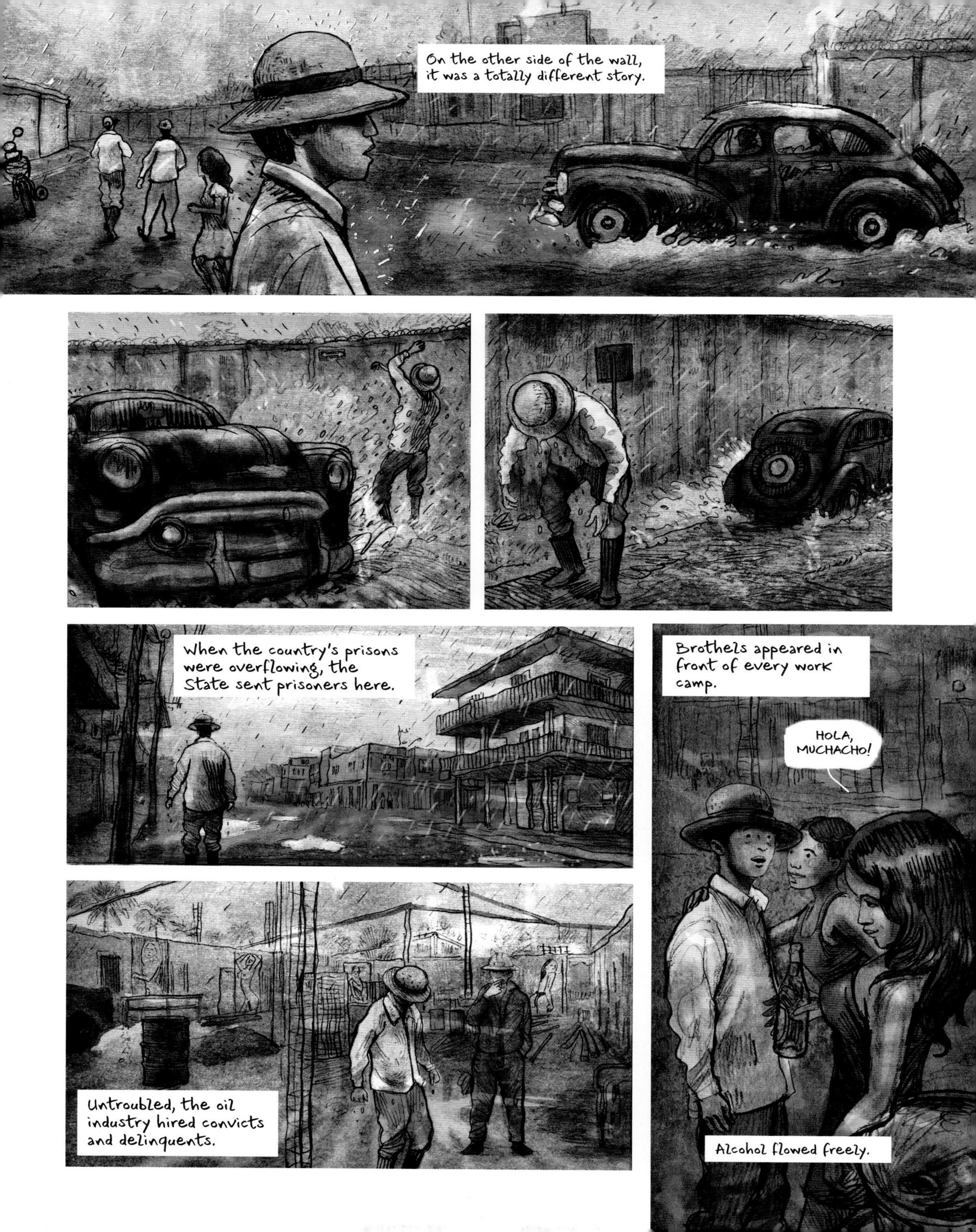
On the other side of the wall, it was a totally different story.
When the country's prisons were overflowing, the State sent prisoners here.
Untroubled, the oil industry hired convicts and delinquents.
Brothels appeared in front of every work camp.
HOLA, MUCHACHO!
Alcohol flowed freely.

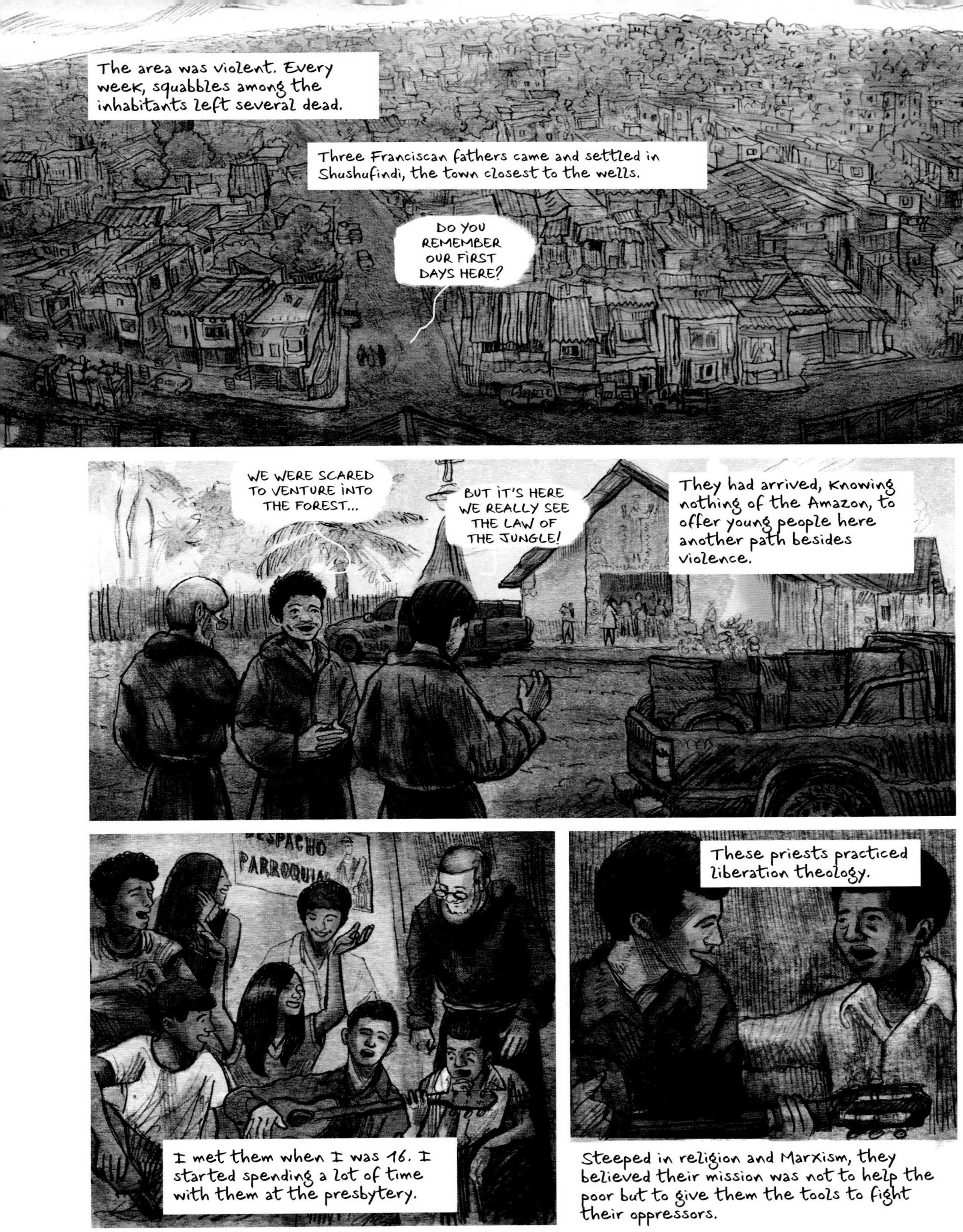
The area was violent. Every week, squabbles among the inhabitants left several dead.
Three Franciscan fathers came and settled in Shushufindi, the town closest to the wells.
DO YOU REMEMBER OUR FIRST DAYS HERE?
WE WERE SCARED TO VENTURE INTO THE FOREST...
BUT IT'S HERE WE REALLY SEE THE LAW OF THE JUNGLE!
They had arrived, knowing nothing of the Amazon, to offer young people here another path besides violence.
DESPACHO PARROQUIAL
I met them when I was 16. I started spending a lot of time with them at the presbytery.
These priests practiced liberation theology.
Steeped in religion and Marxism, they believed their mission was not to help the poor but to give them the tools to fight their oppressors.

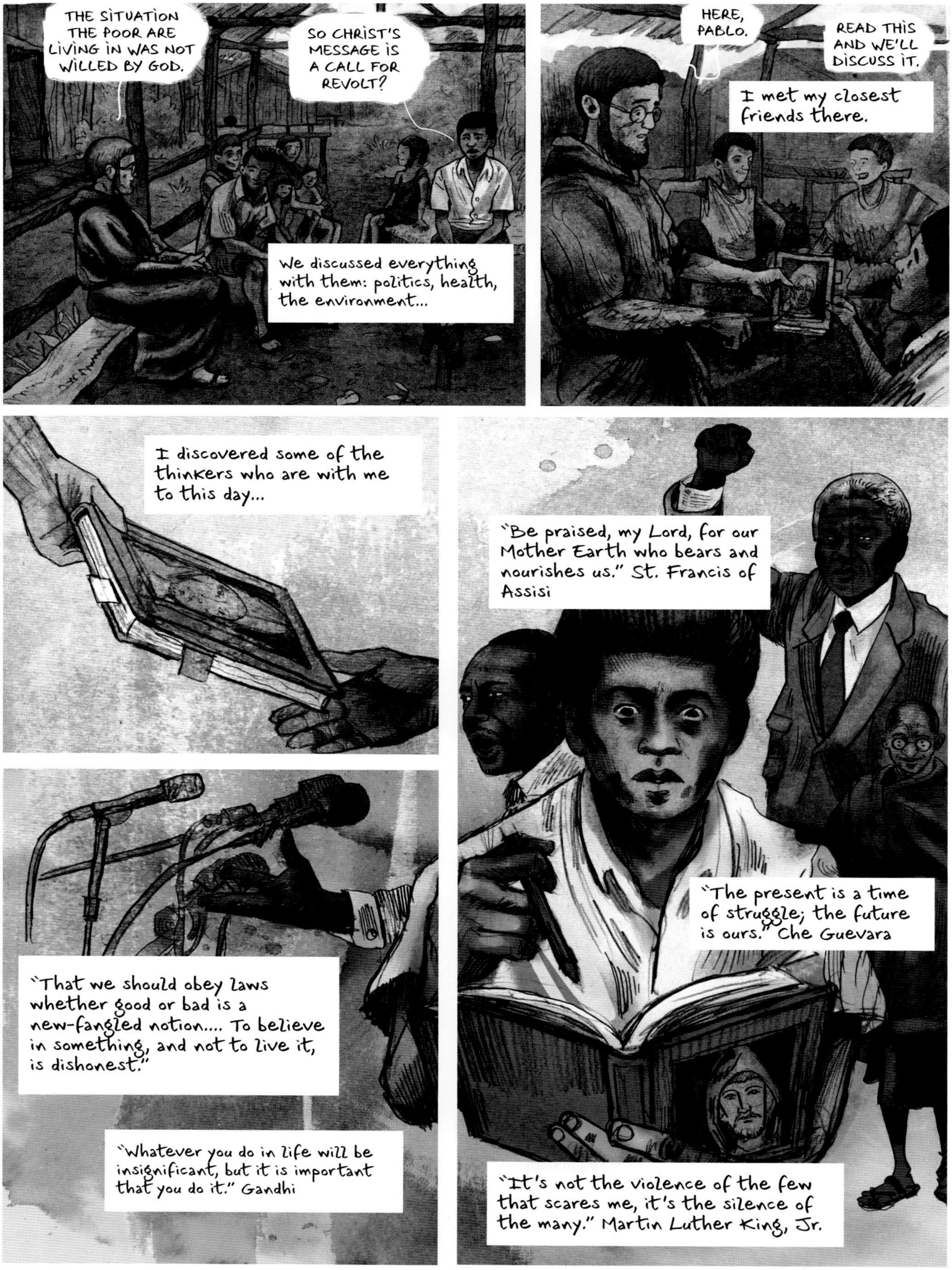
THE SITUATION THE POOR ARE LIVING IN WAS NOT WILLED BY GOD.
SO CHRIST'S MESSAGE IS A CALL FOR REVOLT?
We discussed everything with them: politics, health, the environment...
HERE, PABLO.
READ THIS AND WE'LL DISCUSS IT.
I met my closest friends there.
I discovered some of the thinkers who are with me to this day...
"Be praised, my Lord, for our Mother Earth who bears and nourishes us." St. Francis of Assisi
"The present is a time of struggle; the future is ours." Che Guevara
"That we should obey laws whether good or bad is a new-fangled notion.... To believe in something, and not to live it, is dishonest."
"Whatever you do in life will be insignificant, but it is important that you do it." Gandhi
"It's not the violence of the few that scares me, it's the silence of the many." Martin Luther King, Jr.

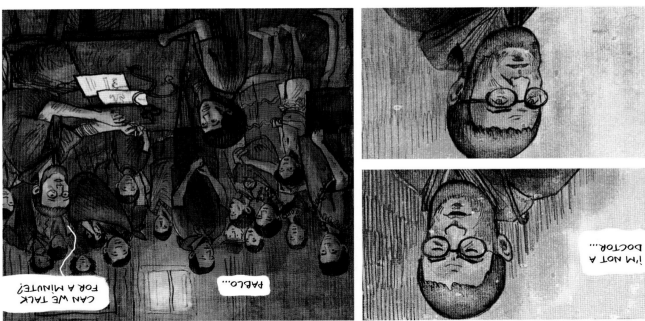

I'M NOT A DOCTOR....

CAN WE TALK FOR A MINUTE!

PABLO....

LISTEN TO HER BREATHING!

WE DON'T KNOW WHAT'S WRONG....

OUR DAUGHTER IS VERY ILL, FATHER.

The Franciscan fathers swiftly became key partners with a suffering people.

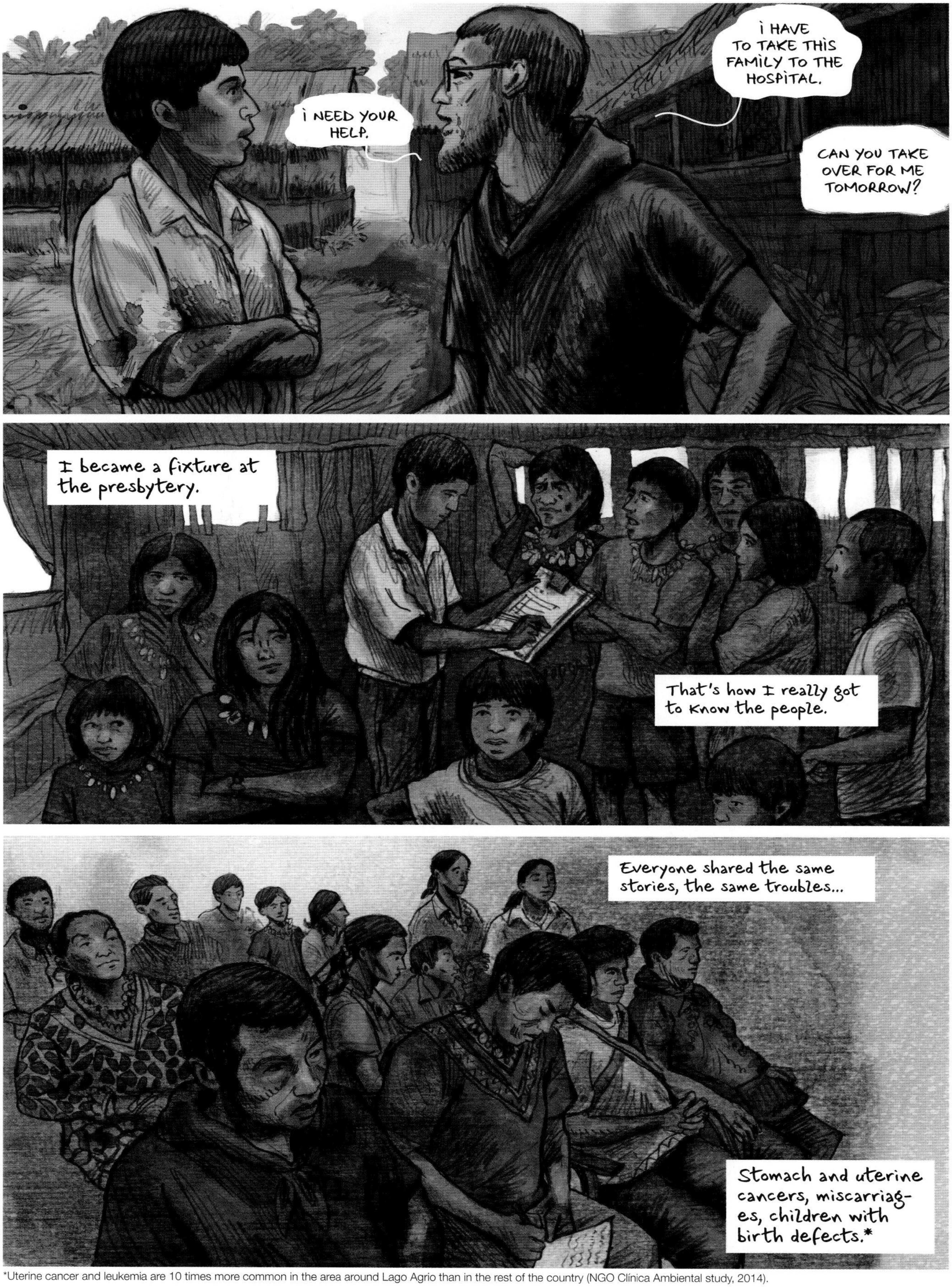
I NEED YOUR HELP.
I HAVE TO TAKE THIS FAMILY TO THE HOSPITAL.
CAN YOU TAKE OVER FOR ME TOMORROW?
I became a fixture at the presbytery.
That's how I really got to know the people.
Everyone shared the same stories, the same troubles...
Stomach and uterine cancers, miscarriag-es, children with birth defects.*

*Uterine cancer and leukemia are 10 times more common in the area around Lago Agrio than in the rest of the country (NGO Clínica Ambiental study, 2014).

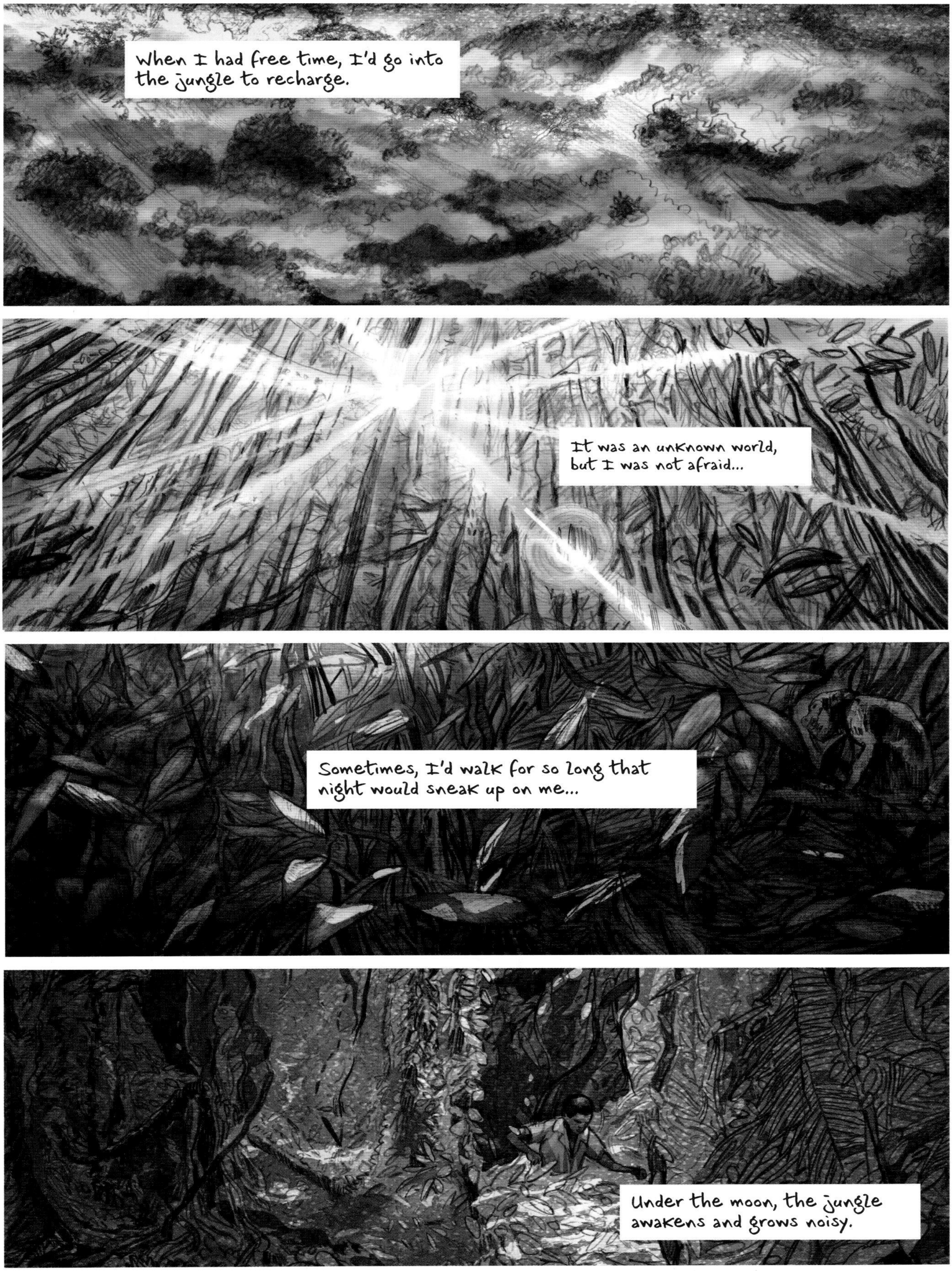
When I had free time, I'd go into the jungle to recharge.
It was an unknown world, but I was not afraid...
Sometimes, I'd walk for so long that night would sneak up on me...
Under the moon, the jungle awakens and grows noisy.

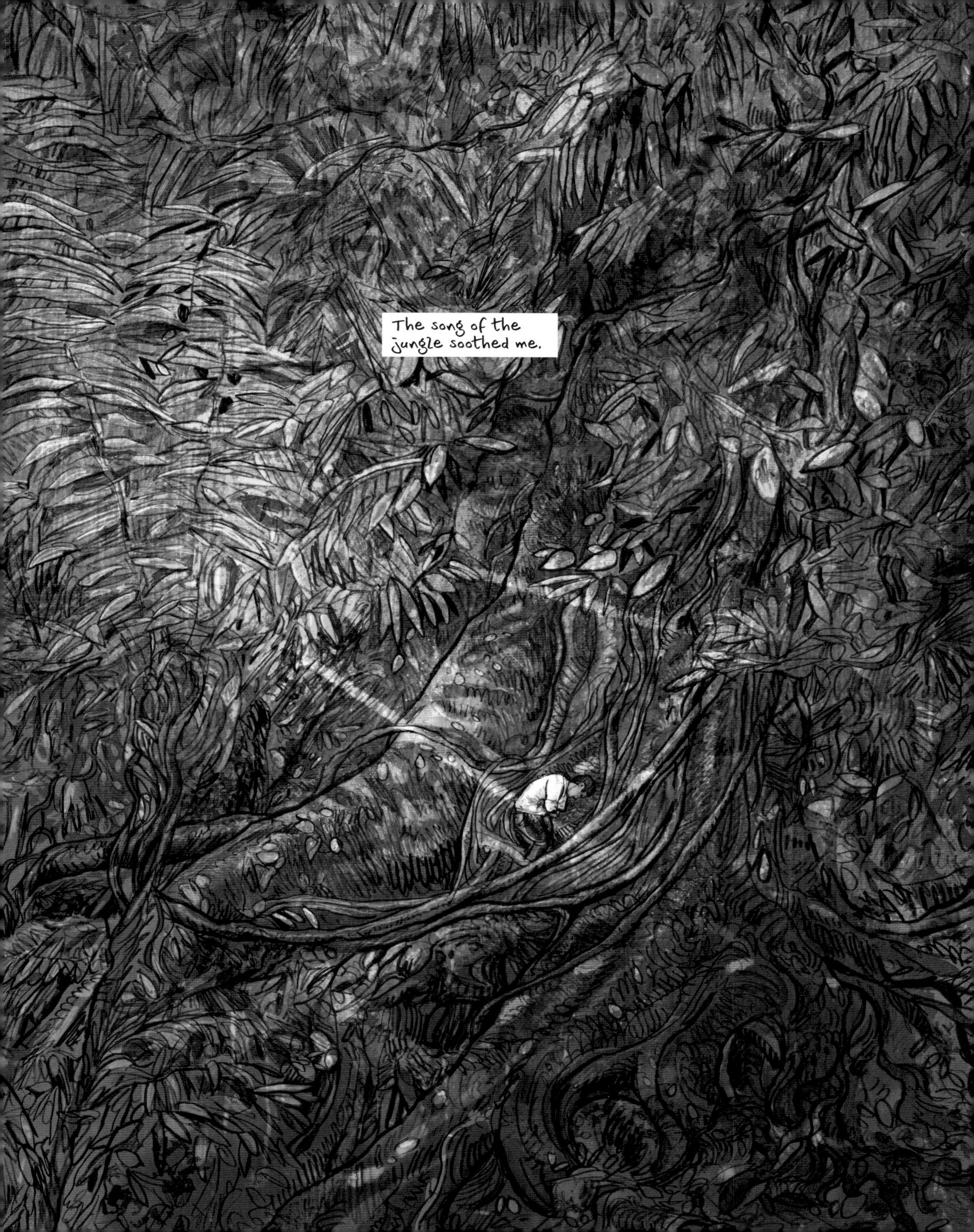
The song of the jungle soothed me.

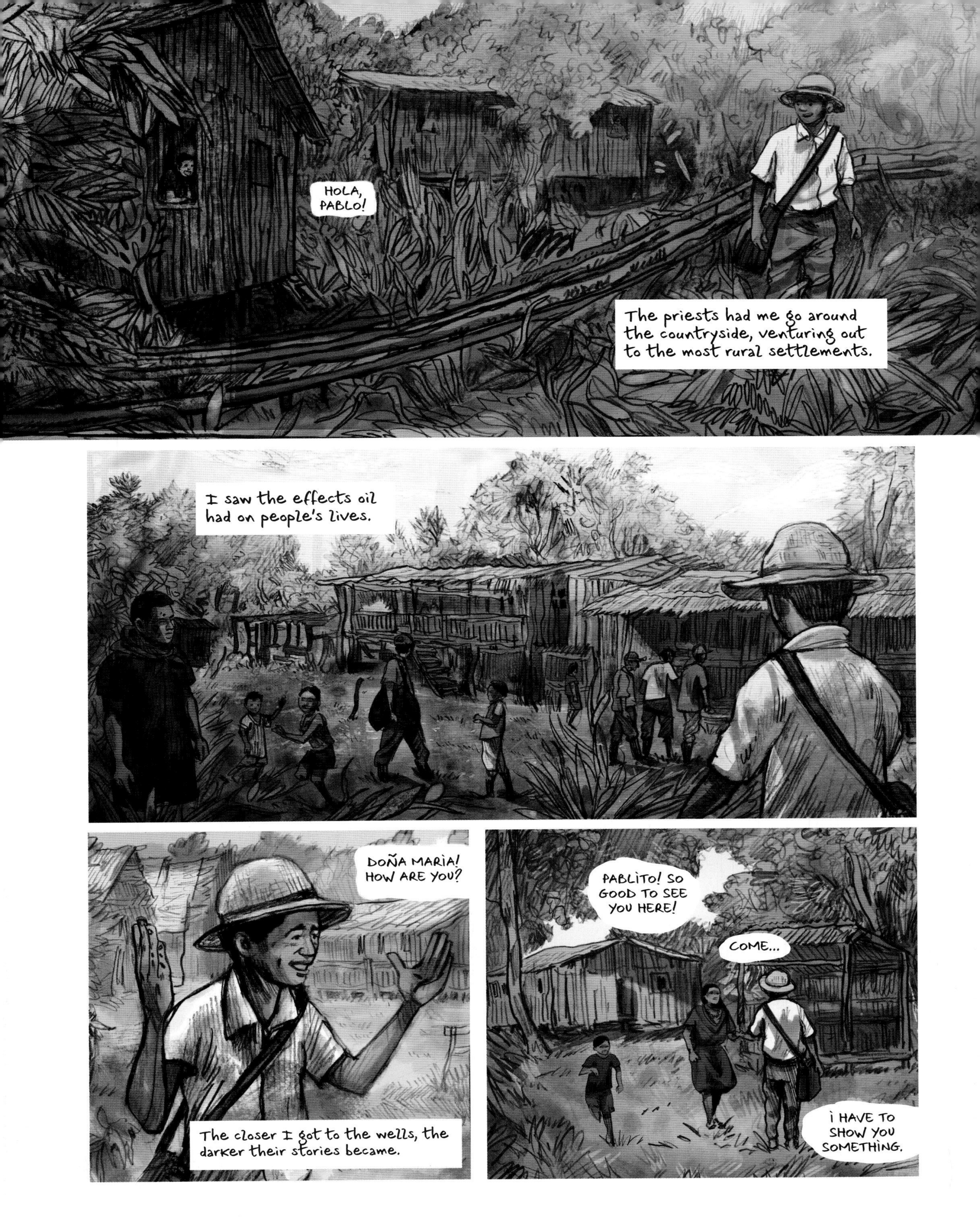
HOLA, PABLO!
The priests had me go around the countryside, venturing out to the most rural settlements.
I saw the effects oil had on people's lives.
DOÑA MARIA! HOW ARE YOU?
The closer I got to the wells, the darker their stories became.
PABLITO! SO GOOD TO SEE YOU HERE!
COME...
¡I HAVE TO SHOW YOU SOMETHING.

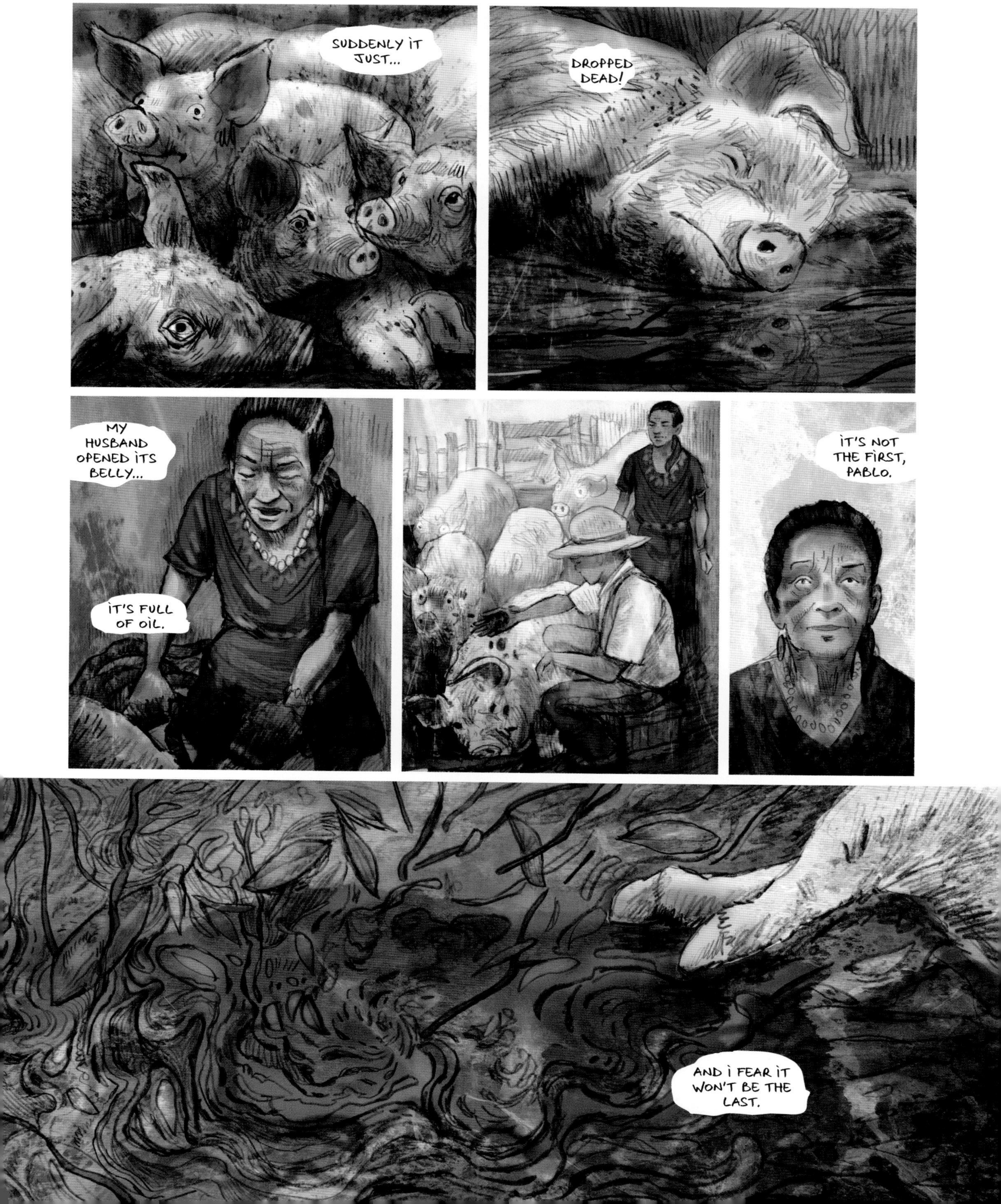
SUDDENLY IT JUST...
DROPPED DEAD!
MY HUSBAND OPENED ITS BELLY...
IT'S FULL OF OIL.
IT'S NOT THE FIRST, PABLO.
AND I FEAR IT WON'T BE THE LAST.

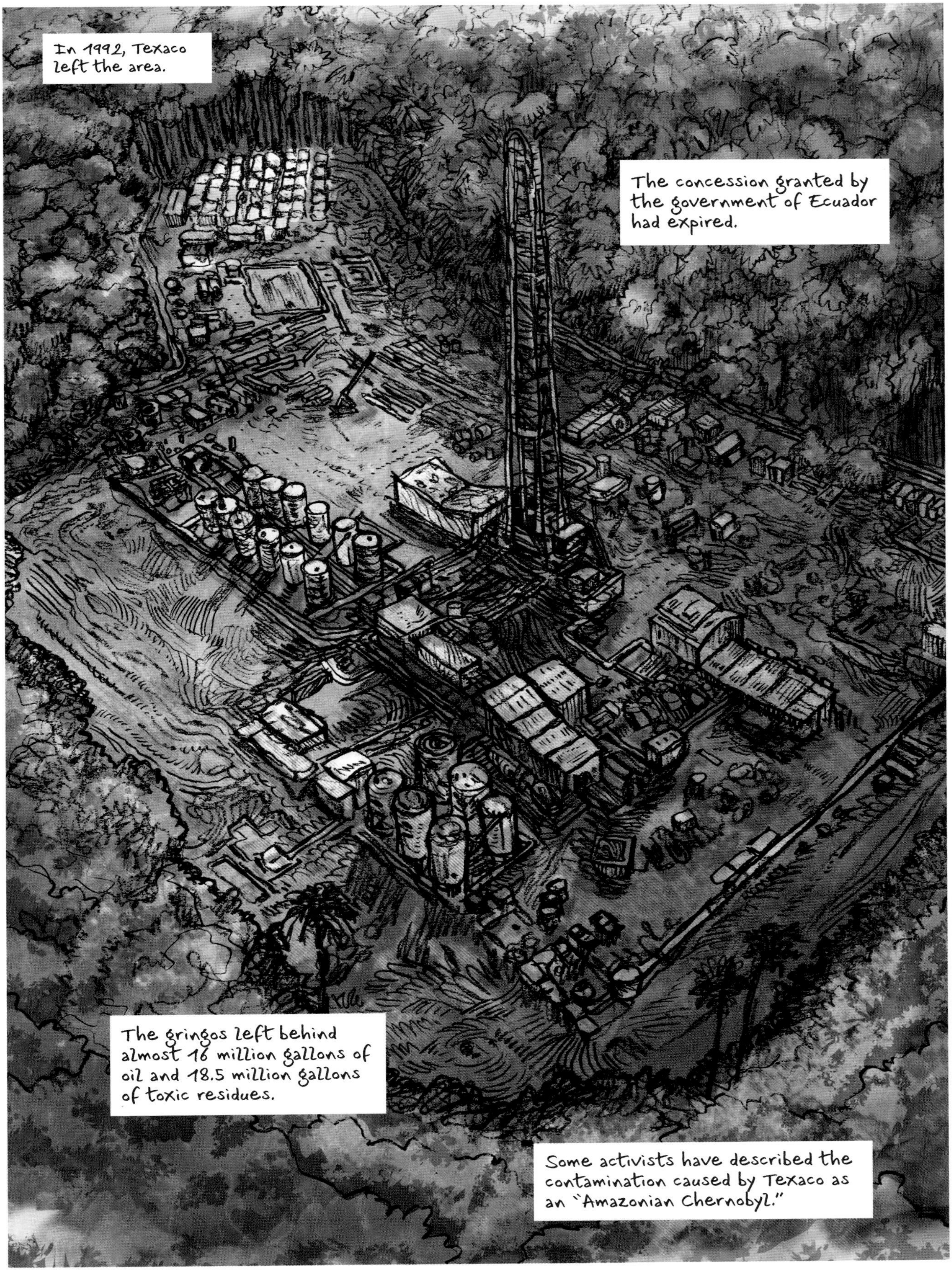
In 1992, Texaco left the area.
The concession granted by the government of Ecuador had expired.
The gringos left behind almost 16 million gallons of oil and 18.5 million gallons of toxic residues.
Some activists have described the contamination caused by Texaco as an "Amazonian Chernobyl."

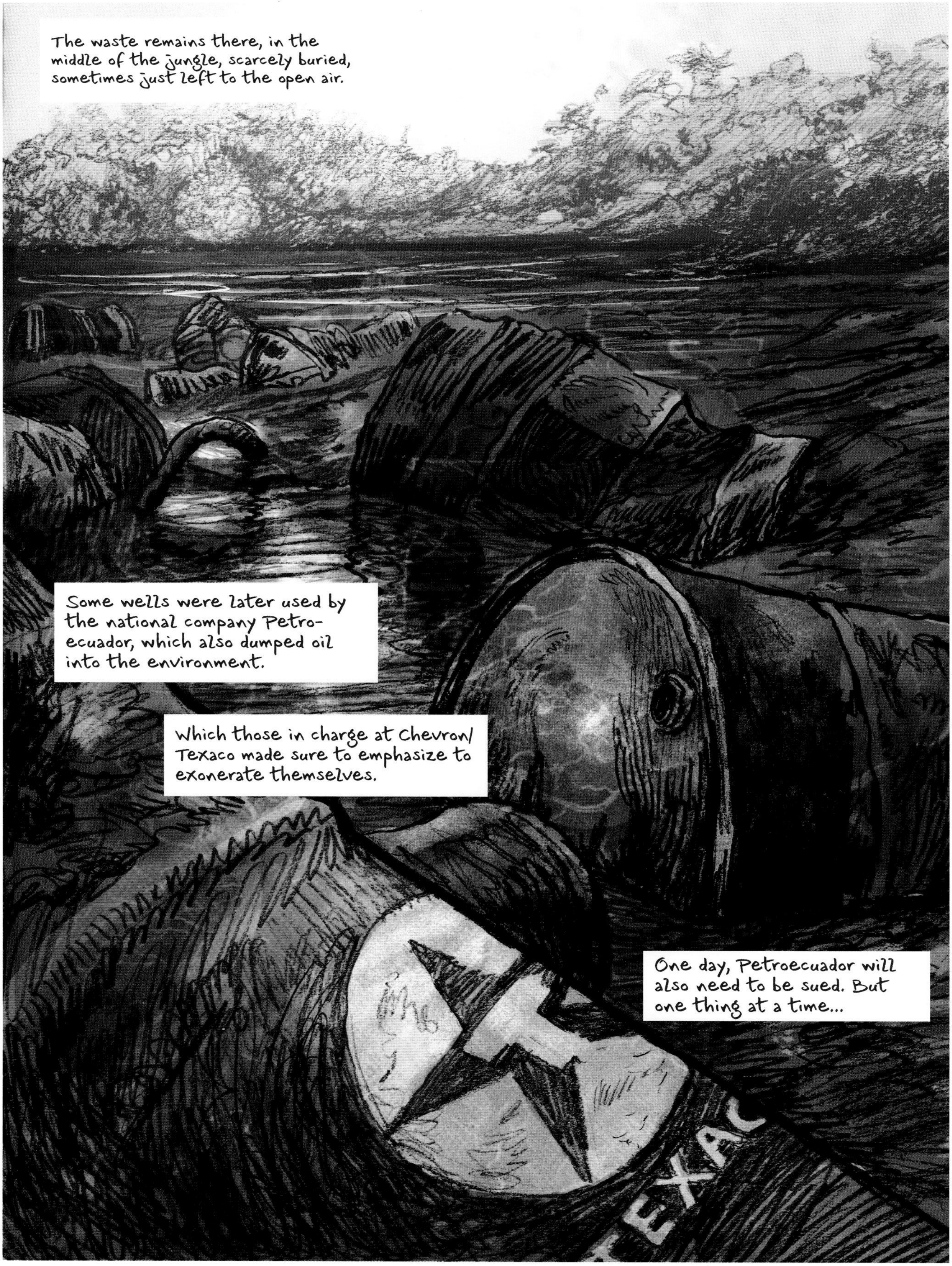
The waste remains there, in the middle of the jungle, scarcely buried, sometimes just left to the open air.
Some wells were later used by the national company Petro-ecuador, which also dumped oil into the environment.
Which those in charge at Chevron/Texaco made sure to emphasize to exonerate themselves.
One day, Petroecuador will also need to be sued. But one thing at a time...
TEXAC

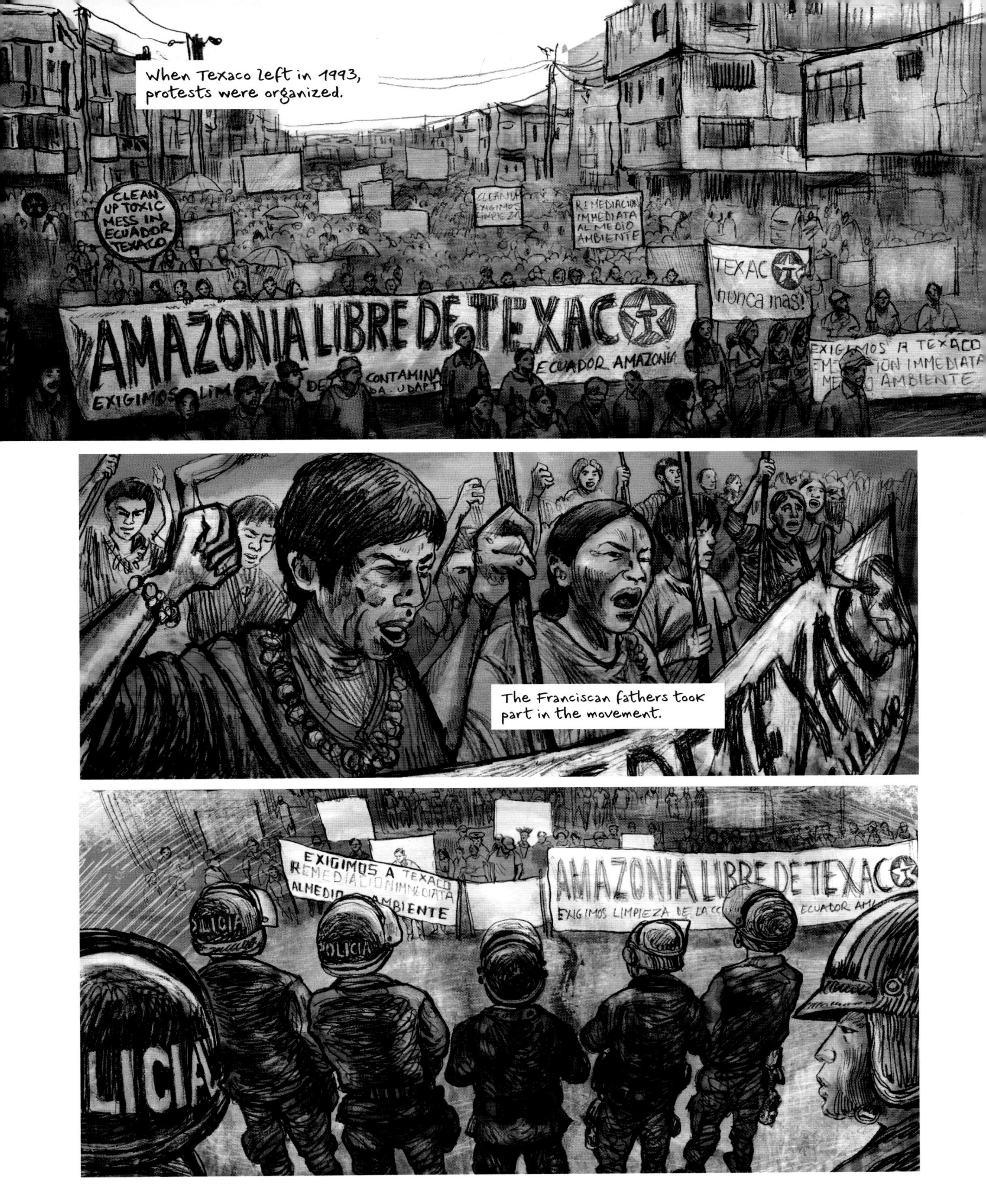
When Texaco left in 1993, protests were organized.
CLEAN UP TOXIC MESS IN ECUADOR TEXACO
REMEDIACION IMMEDIATA AL MEDIO AMBIENTE
AMAZONIA LIBRE DE TEXACO
ECUADOR AMAZONIA
TEXACO nunca mas!
The Franciscan fathers took part in the movement.
EXIGIMOS A TEXACO REMEDIACION IMMEDIATA ALMEDIO AMBIENTE
AMAZONIA LIBRE DE TEXACO
POLICIA

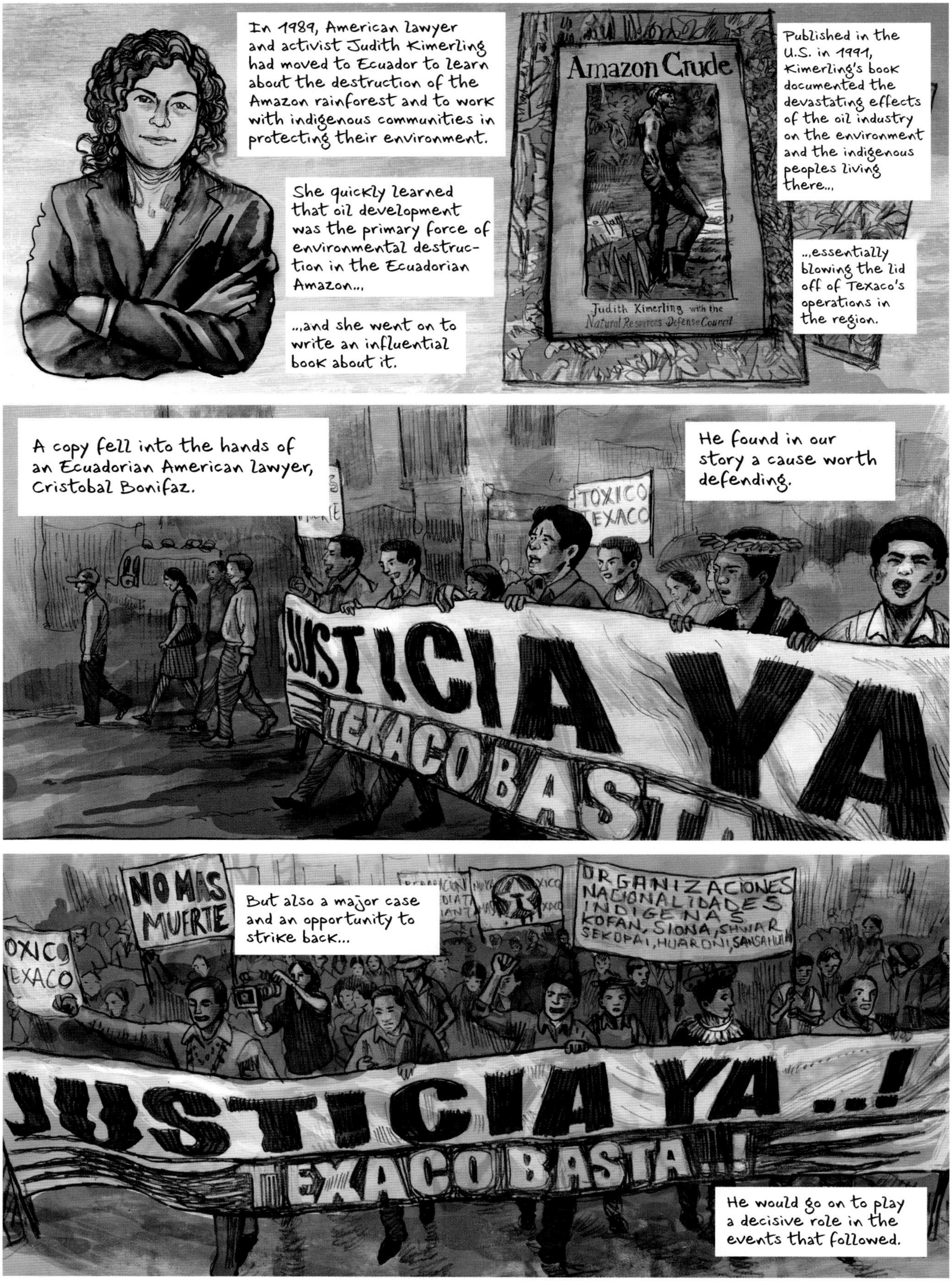
In 1989, American lawyer and activist Judith Kimerling had moved to Ecuador to learn about the destruction of the Amazon rainforest and to work with indigenous communities in protecting their environment.
She quickly learned that oil development was the primary force of environmental destruction in the Ecuadorian Amazon...
...and she went on to write an influential book about it.
Amazon Crude
Judith Kimerling with the Natural Resources Defense Council
Published in the U.S. in 1991, Kimerling's book documented the devastating effects of the oil industry on the environment and the indigenous peoples living there...
...essentially blowing the lid off of Texaco's operations in the region.
A copy fell into the hands of an Ecuadorian American lawyer, Cristobal Bonifaz.
He found in our story a cause worth defending.
TOXICO TEXACO
JUSTICIA YA
TEXACO BASTA
NO MAS MUERTE
But also a major case and an opportunity to strike back...
ORGANIZACIONES NACIONALIDADES INDIGENAS
KOFAN, SIONA, SHWAR
JUSTICIA YA..!
TEXACO BASTA..!
He would go on to play a decisive role in the events that followed.

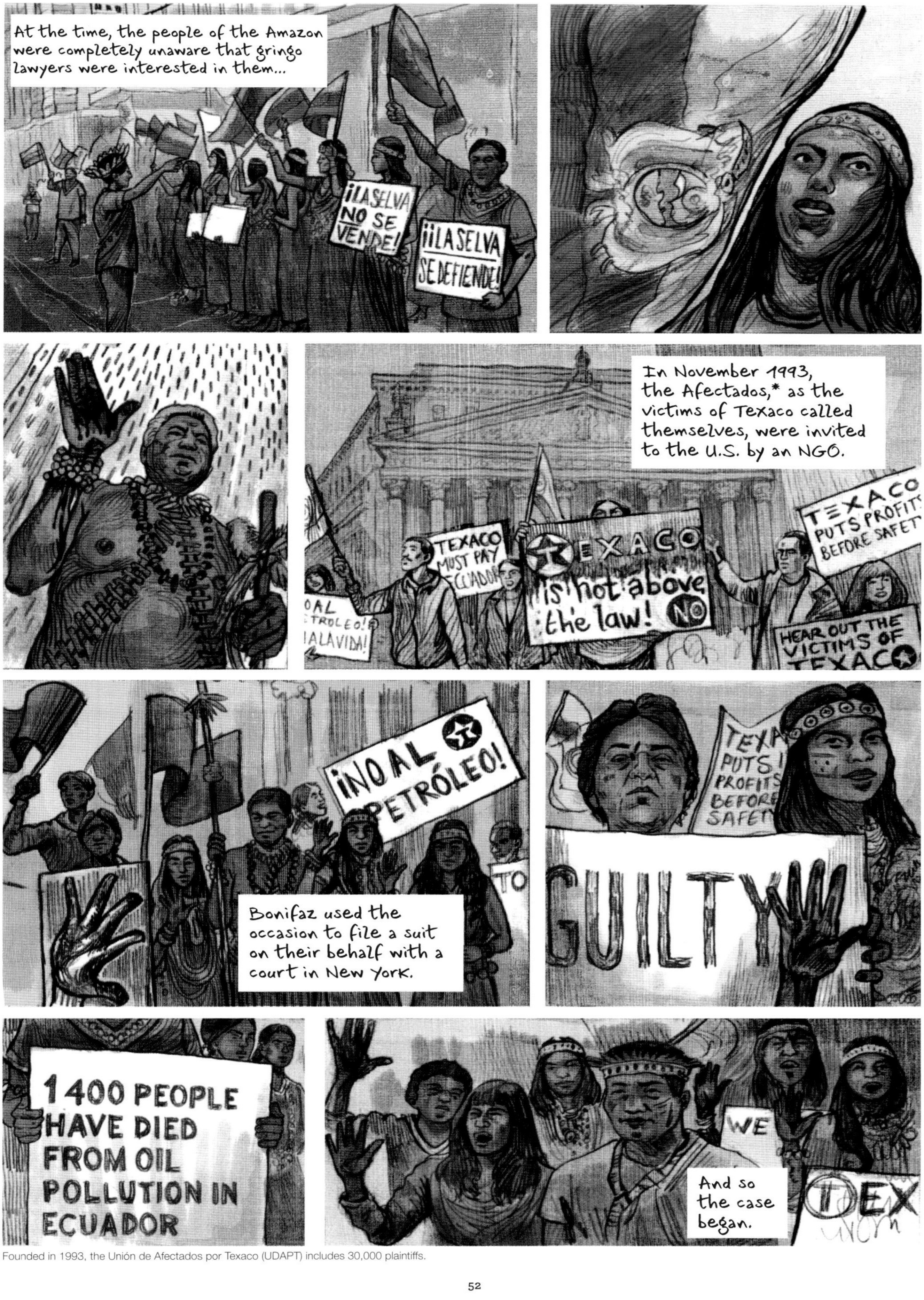
At the time, the people of the Amazon were completely unaware that gringo lawyers were interested in them...
¡¡LA SELVA NO SE VENDE!
¡¡LA SELVA SE DEFIENDE!
In November 1993, the Afectados,* as the victims of Texaco called themselves, were invited to the U.S. by an NGO.
TEXACO MUST PAY
TEXACO is not above the law! NO
TEXACO PUTS PROFITS BEFORE SAFETY
HEAR OUT THE VICTIMS OF TEXACO
¡NO AL PETRÓLEO!
Bonifaz used the occasion to file a suit on their behalf with a court in New York.
PUTS PROFITS BEFORE SAFETY
GUILTY
1400 PEOPLE HAVE DIED FROM OIL POLLUTION IN ECUADOR
WE
And so the case began.

Founded in 1993, the Unión de Afectados por Texaco (UDAPT) includes 30,000 plaintiffs.

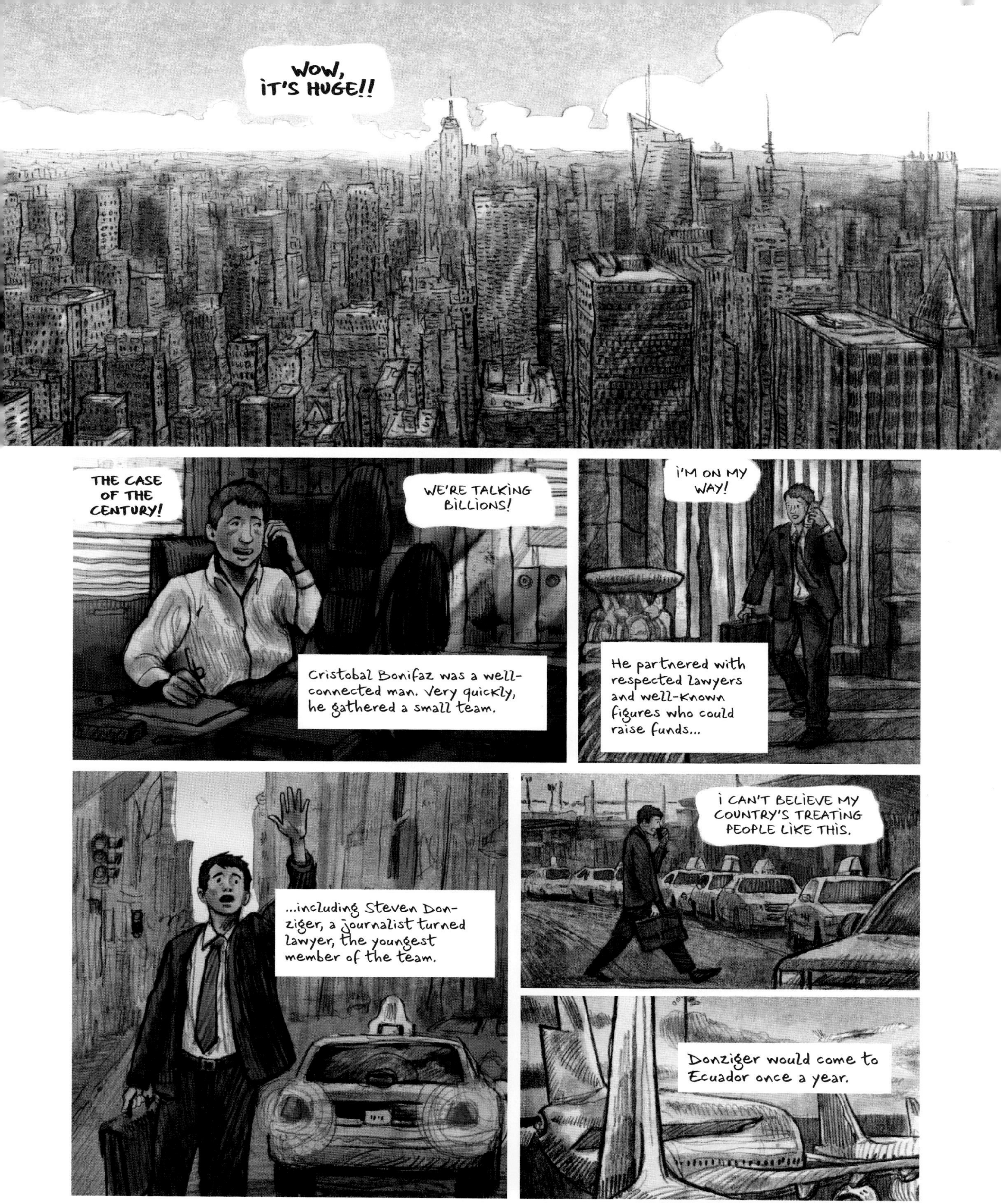
WOW, IT'S HUGE!!
THE CASE OF THE CENTURY!
WE'RE TALKING BILLIONS!
Cristobal Bonifaz was a well-connected man. Very quickly, he gathered a small team.
I'M ON MY WAY!
He partnered with respected lawyers and well-known figures who could raise funds...
...including Steven Donziger, a journalist turned lawyer, the youngest member of the team.
I CAN'T BELIEVE MY COUNTRY'S TREATING PEOPLE LIKE THIS.
Donziger would come to Ecuador once a year.

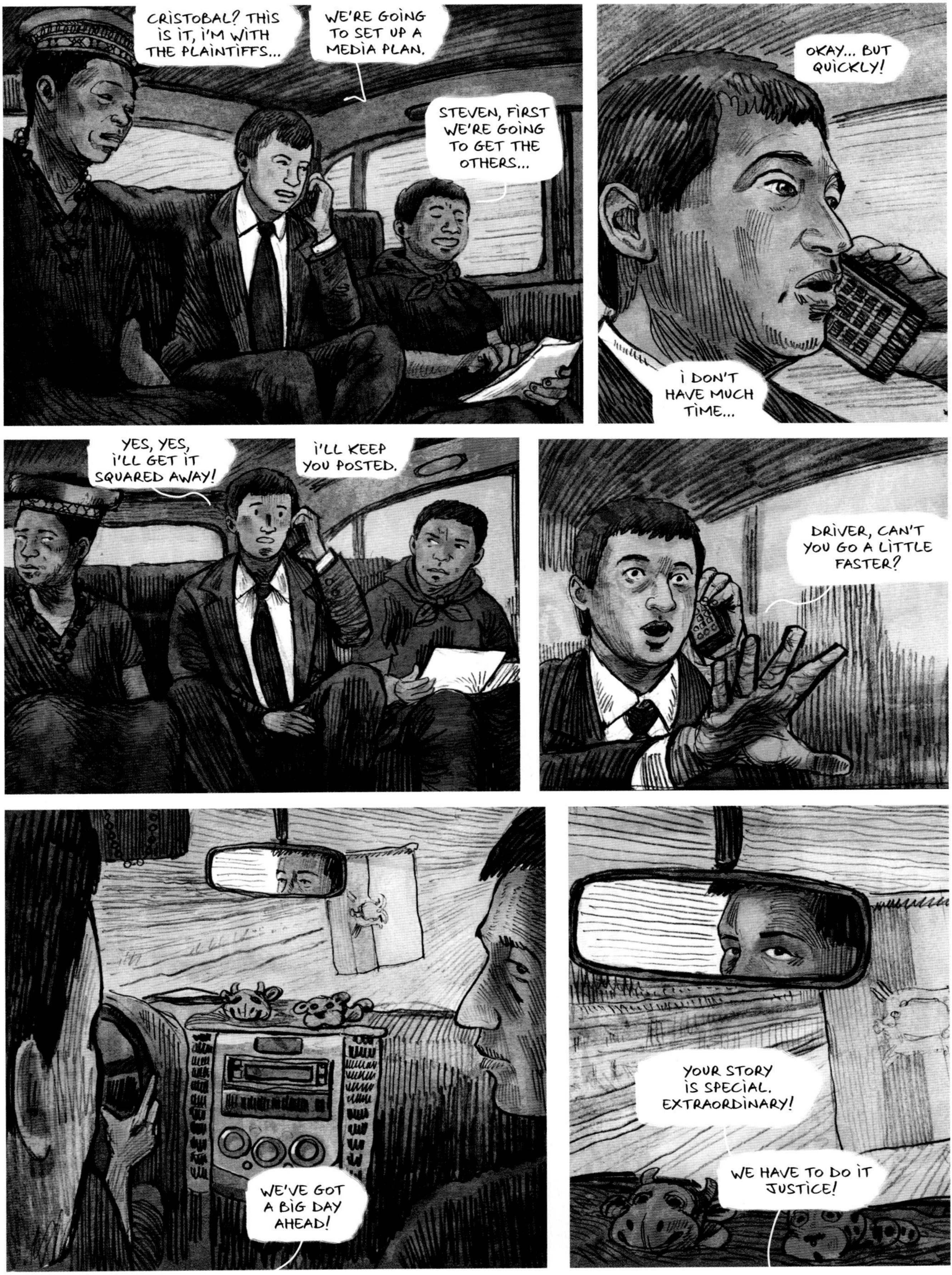
CRISTOBAL? THIS IS IT, I'M WITH THE PLAINTIFFS...
WE'RE GOING TO SET UP A MEDIA PLAN.
STEVEN, FIRST WE'RE GOING TO GET THE OTHERS...
OKAY... BUT QUICKLY!
I DON'T HAVE MUCH TIME...
YES, YES, I'LL GET IT SQUARED AWAY!
I'LL KEEP YOU POSTED.
DRIVER, CAN'T YOU GO A LITTLE FASTER?
WE'VE GOT A BIG DAY AHEAD!
YOUR STORY IS SPECIAL. EXTRAORDINARY!
WE HAVE TO DO IT JUSTICE!

On December 7, 1996, the opening hearing took place in New York. Chevron called for the charges to be dropped. Wisely, the company also requested that, if there must be a trial, it should be in Ecuador.

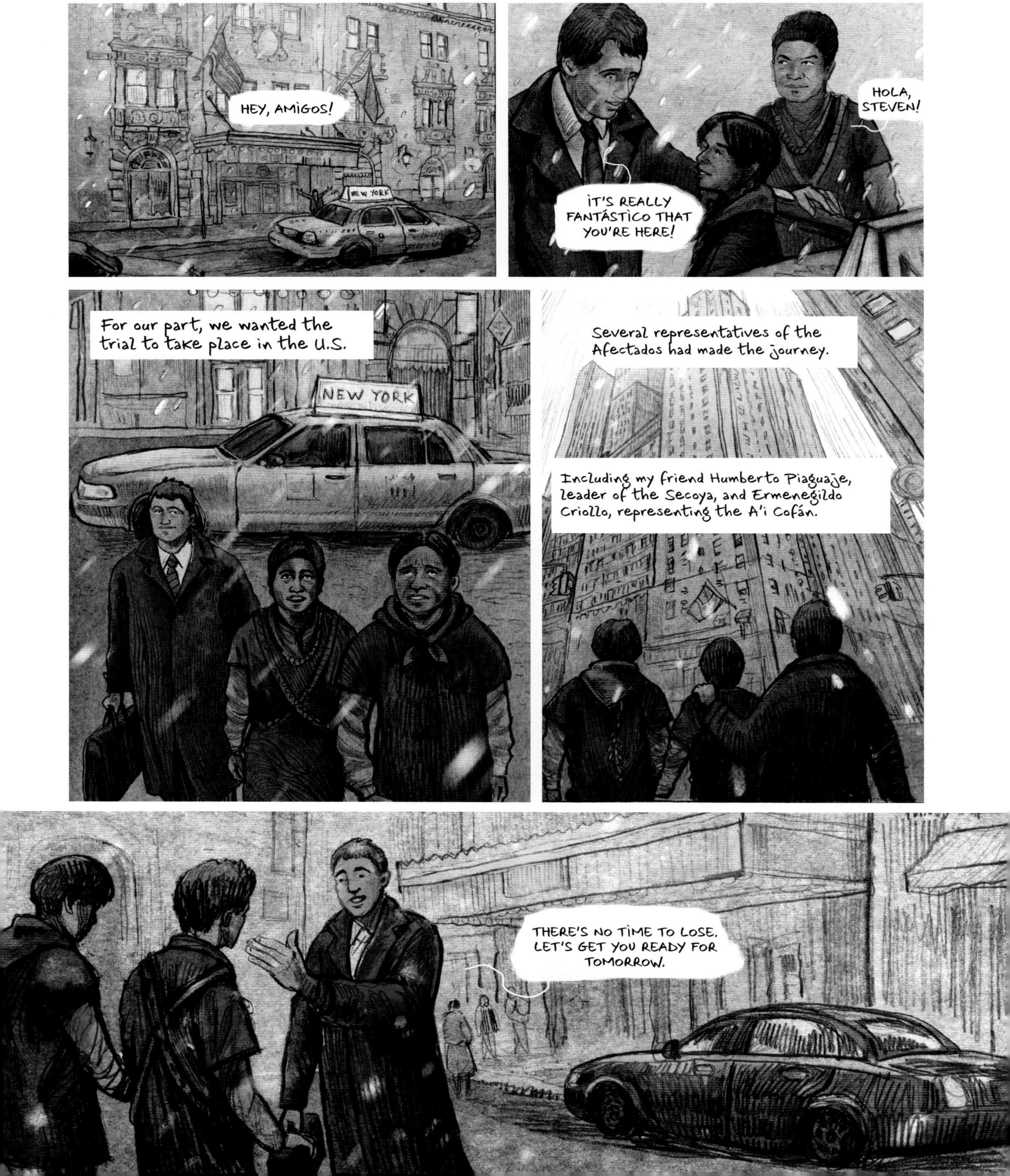
HEY, AMIGOS!
HOLA, STEVEN!
IT'S REALLY FANTÁSTICO THAT YOU'RE HERE!
For our part, we wanted the trial to take place in the U.S.
NEW YORK
Several representatives of the Afectados had made the journey.
Including my friend Humberto Piaguaje, leader of the Secoya, and Ermenegildo Criollo, representing the A'i Cofán.
THERE'S NO TIME TO LOSE. LET'S GET YOU READY FOR TOMORROW.

TELL ME A BIT ABOUT WHAT YOU'RE GOING TO SAY TO THE JUDGE.
WELL...
OUR FOREST IS THE LUNGS OF THE PLANET...
...AND YOU'RE SLAUGH-TERING IT!
THEN WE'LL SHARE THE NUMBERS...
HMM...
OKAY, IT'S NOT BAD...
BUT IT NEEDS MORE EMOTION!
I HAVE SOME IDEAS FOR YOU. WRITE THESE DOWN.
YOUR HONOR, I'VE TRAVELED A LONG WAY TO COME BEFORE YOU...
I JOURNEYED BY CANOE, BUS, AND PLANE...
YOUR COUNTRY SPENT 28 YEARS ON MY LAND...
SURELY YOU CAN GRANT ME 3 MINUTES ON YOURS!
YOU'LL SAY ALL THAT IN YOUR TRADITIONAL GARB. THAT'LL GET THEM GOING FOR US!
I'LL LET YOU PRACTICE. SEE YOU TOMORROW!

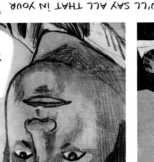

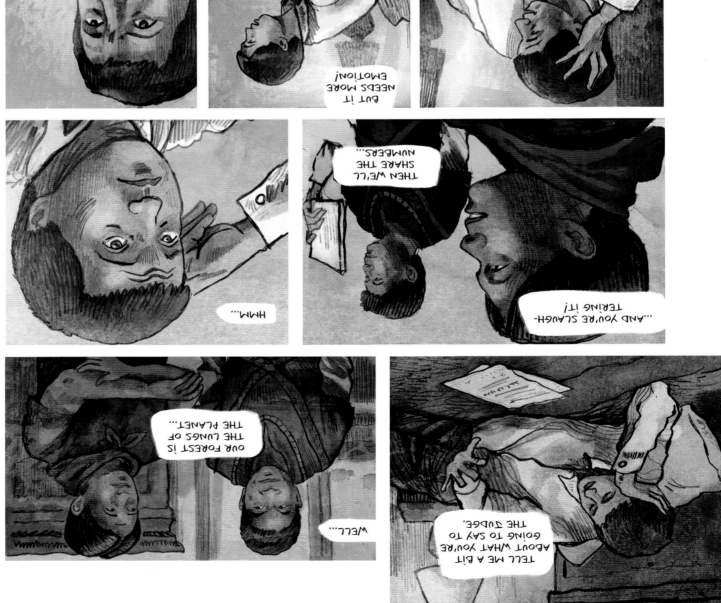

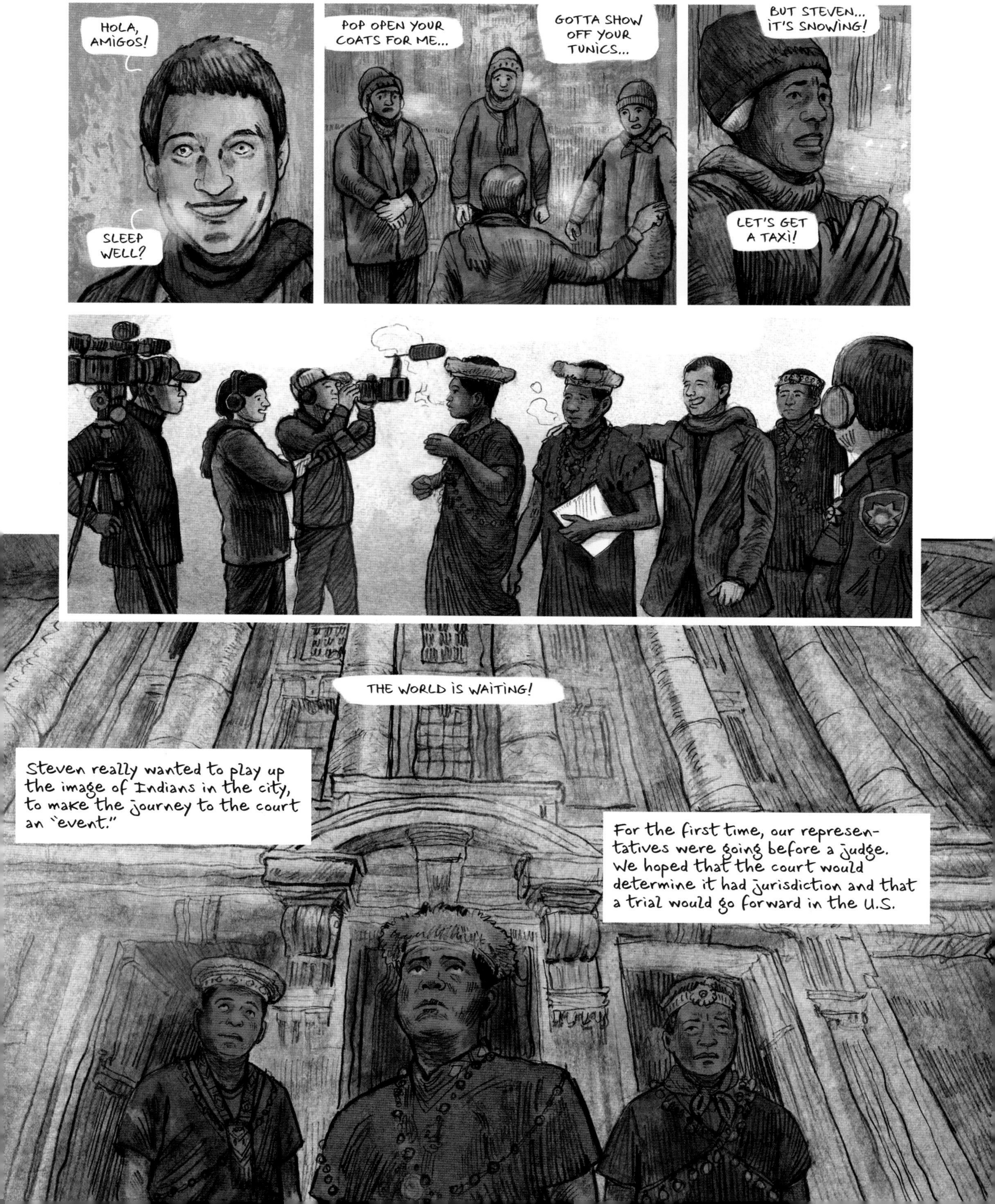
HOLA, AMIGOS!
SLEEP WELL?
POP OPEN YOUR COATS FOR ME...
GOTTA SHOW OFF YOUR TUNICS...
BUT STEVEN... IT'S SNOWING!
LET'S GET A TAXI!
THE WORLD IS WAITING!
Steven really wanted to play up the image of Indians in the city, to make the journey to the court an "event."
For the first time, our representatives were going before a judge. We hoped that the court would determine it had jurisdiction and that a trial would go forward in the U.S.

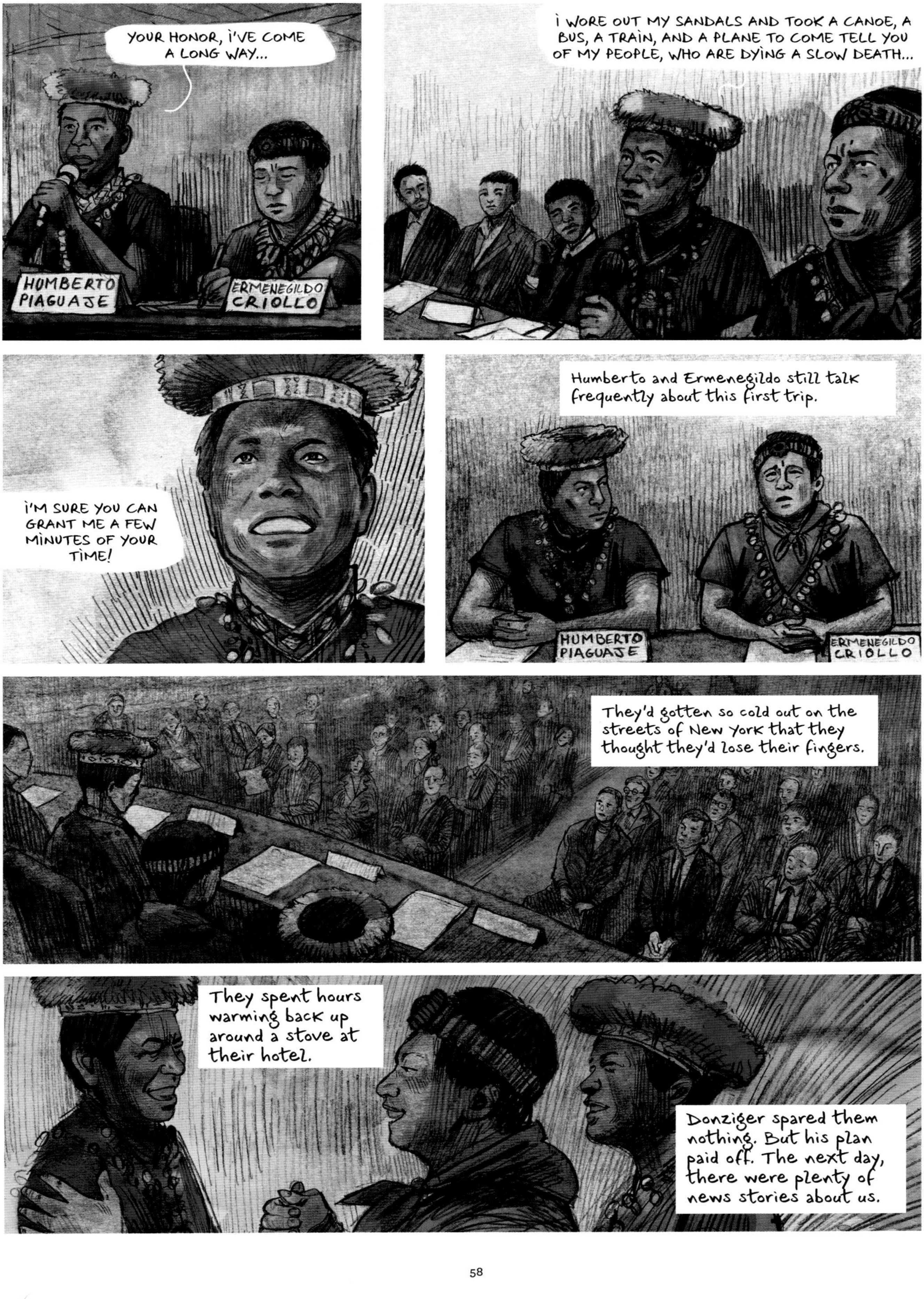
YOUR HONOR, I'VE COME A LONG WAY...
HUMBERTO PIAGUAJE
ERMENEGILDO CRIOLLO
I WORE OUT MY SANDALS AND TOOK A CANOE, A BUS, A TRAIN, AND A PLANE TO COME TELL YOU OF MY PEOPLE, WHO ARE DYING A SLOW DEATH...
I'M SURE YOU CAN GRANT ME A FEW MINUTES OF YOUR TIME!
Humberto and Ermenegildo still talk frequently about this first trip.
HUMBERTO PIAGUAJE
ERMENEGILDO CRIOLLO
They'd gotten so cold out on the streets of New York that they thought they'd lose their fingers.
They spent hours warming back up around a stove at their hotel.
Donziger spared them nothing. But his plan paid off. The next day, there were plenty of news stories about us.

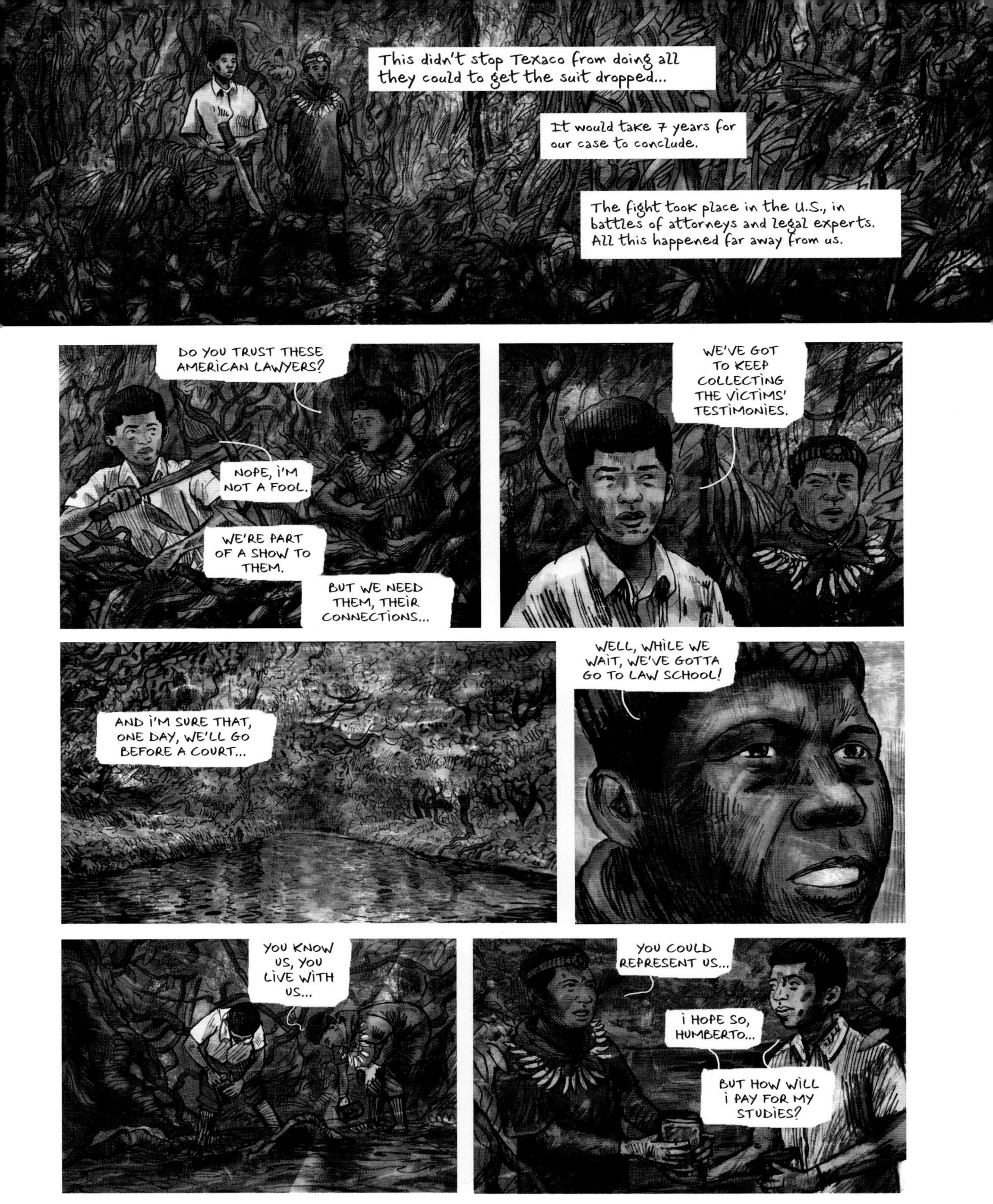
This didn't stop Texaco from doing all they could to get the suit dropped...
It would take 7 years for our case to conclude.
The fight took place in the U.S., in battles of attorneys and legal experts. All this happened far away from us.
DO YOU TRUST THESE AMERICAN LAWYERS?
NOPE, I'M NOT A FOOL.
WE'RE PART OF A SHOW TO THEM.
BUT WE NEED THEM, THEIR CONNECTIONS...
WE'VE GOT TO KEEP COLLECTING THE VICTIMS' TESTIMONIES.
AND I'M SURE THAT, ONE DAY, WE'LL GO BEFORE A COURT...
WELL, WHILE WE WAIT, WE'VE GOTTA GO TO LAW SCHOOL!
YOU KNOW US, YOU LIVE WITH US...
YOU COULD REPRESENT US...
I HOPE SO, HUMBERTO...
BUT HOW WILL I PAY FOR MY STUDIES?

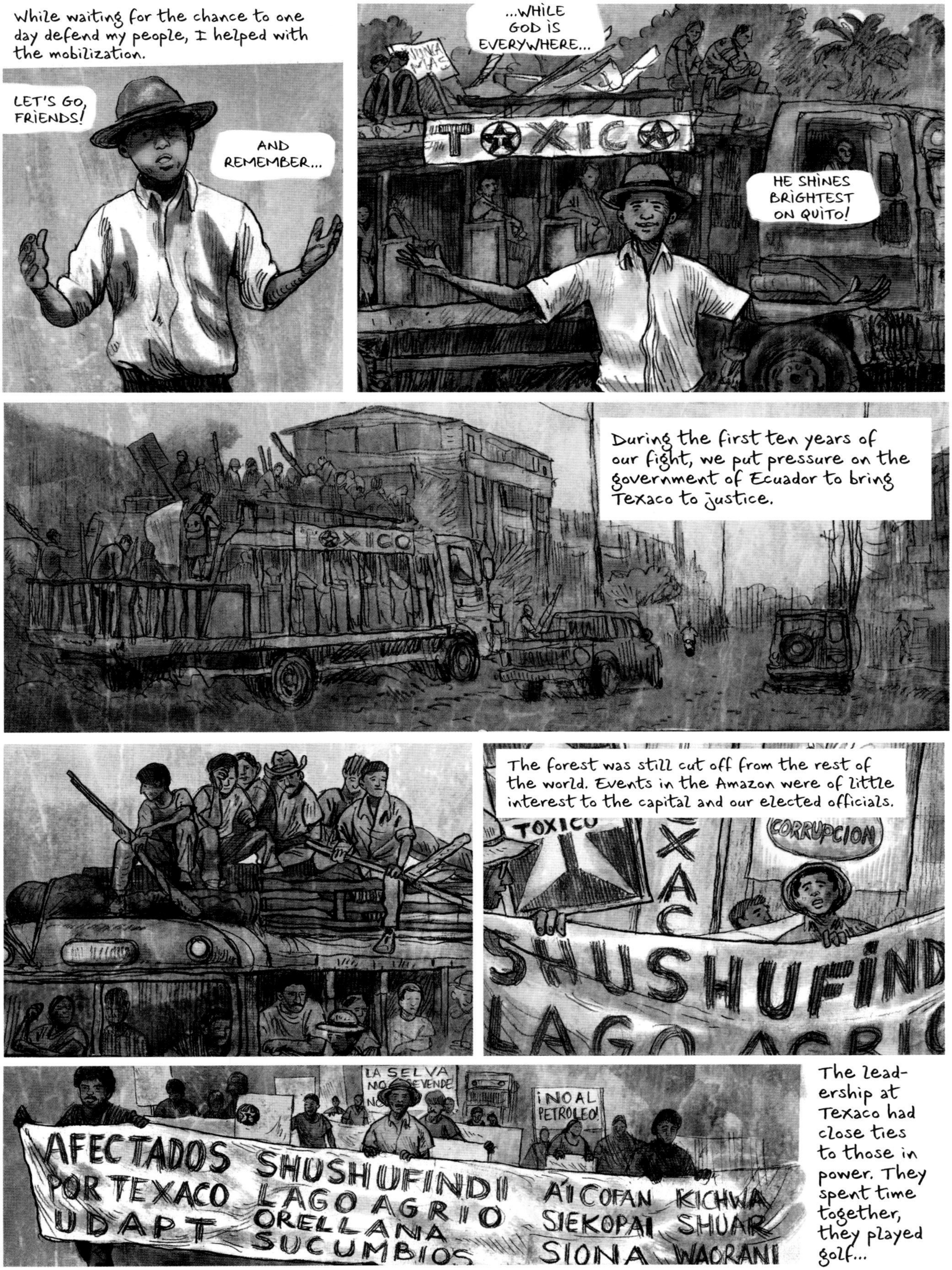
While waiting for the chance to one day defend my people, I helped with the mobilization.
LET'S GO, FRIENDS!
AND REMEMBER...
...WHILE GOD IS EVERYWHERE...
TOXICO
HE SHINES BRIGHTEST ON QUITO!
During the first ten years of our fight, we put pressure on the government of Ecuador to bring Texaco to justice.
TOXICO
The forest was still cut off from the rest of the world. Events in the Amazon were of little interest to the capital and our elected officials.
TOXICO
CORRUPCION
SHUSHUFIND
LAGO AGRI
LA SELVA NO SE VENDE
¡NO AL PETROLEO!
AFECTADOS POR TEXACO UDAPT
SHUSHUFINDI LAGO AGRIO ORELLANA SUCUMBIOS
AÍCOFAN SIEKOPAI SIONA
KICHWA SHUAR WAORANI
The leadership at Texaco had close ties to those in power. They spent time together, they played golf...

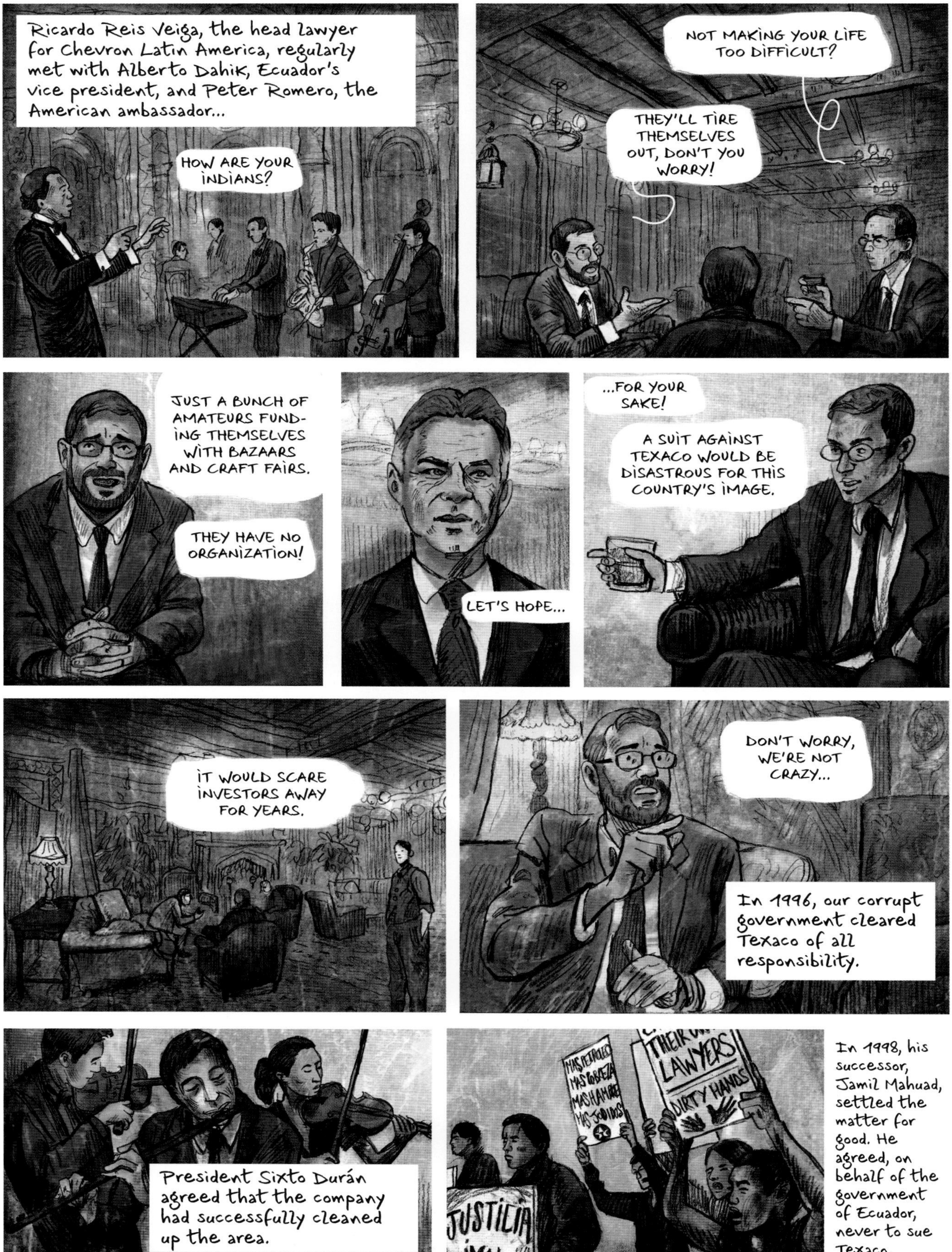
Ricardo Reis Veiga, the head lawyer for Chevron Latin America, regularly met with Alberto Dahik, Ecuador's vice president, and Peter Romero, the American ambassador...
HOW ARE YOUR INDIANS?
NOT MAKING YOUR LIFE TOO DIFFICULT?
THEY'LL TIRE THEMSELVES OUT, DON'T YOU WORRY!
JUST A BUNCH OF AMATEURS FUNDING THEMSELVES WITH BAZAARS AND CRAFT FAIRS.
THEY HAVE NO ORGANIZATION!
LET'S HOPE...
...FOR YOUR SAKE!
A SUIT AGAINST TEXACO WOULD BE DISASTROUS FOR THIS COUNTRY'S IMAGE.
IT WOULD SCARE INVESTORS AWAY FOR YEARS.
DON'T WORRY, WE'RE NOT CRAZY...
In 1996, our corrupt government cleared Texaco of all responsibility.
President Sixto Durán agreed that the company had successfully cleaned up the area.
LAWYERS
DIRTY HANDS
¡YA!
In 1998, his successor, Jamil Mahuad, settled the matter for good. He agreed, on behalf of the government of Ecuador, never to sue Texaco.

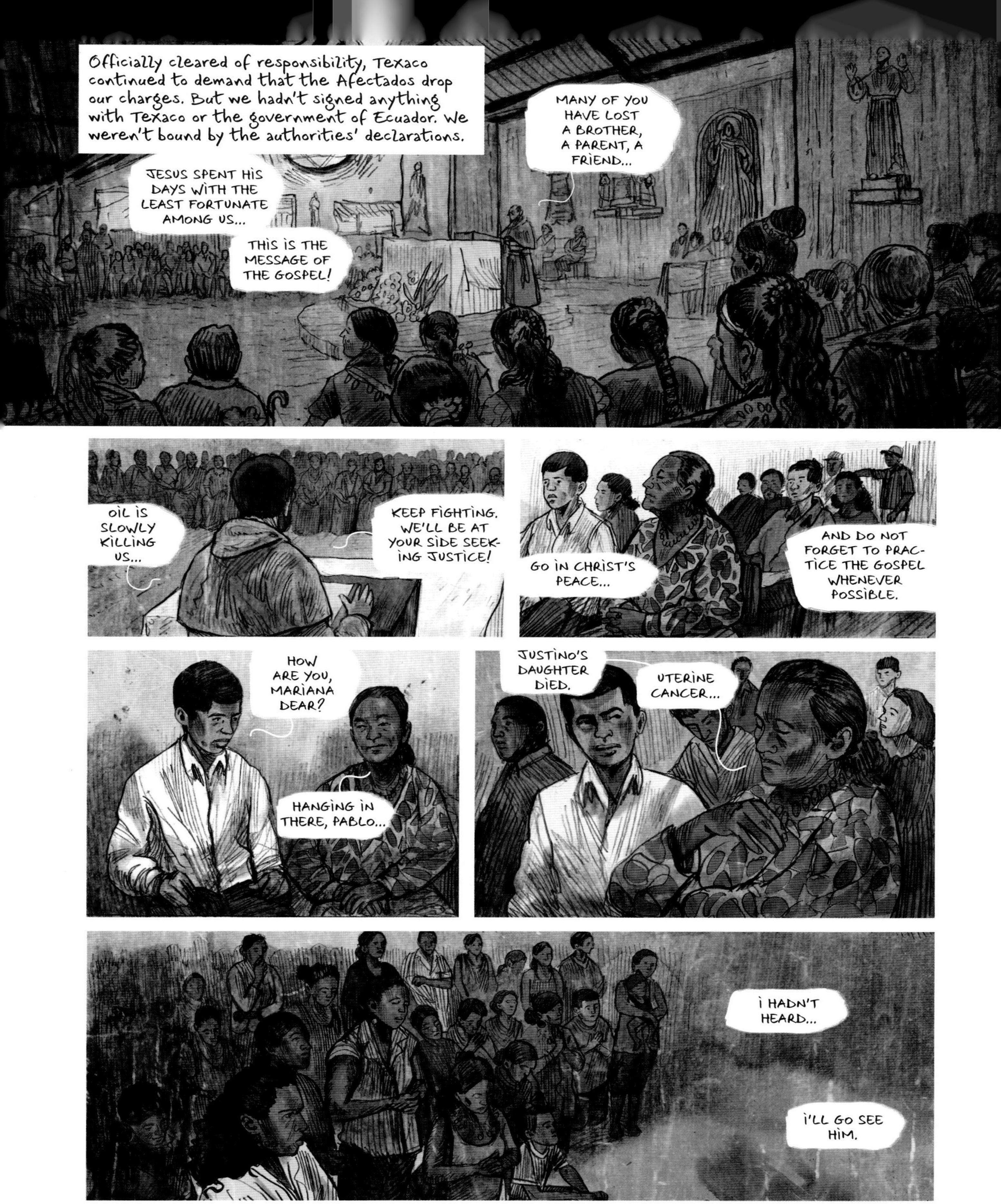
Officially cleared of responsibility, Texaco continued to demand that the Afectados drop our charges. But we hadn't signed anything with Texaco or the government of Ecuador. We weren't bound by the authorities' declarations.
JESUS SPENT HIS DAYS WITH THE LEAST FORTUNATE AMONG US...
THIS IS THE MESSAGE OF THE GOSPEL!
MANY OF YOU HAVE LOST A BROTHER, A PARENT, A FRIEND...
OIL IS SLOWLY KILLING US...
KEEP FIGHTING. WE'LL BE AT YOUR SIDE SEEK-ING JUSTICE!
GO IN CHRIST'S PEACE...
AND DO NOT FORGET TO PRAC-TICE THE GOSPEL WHENEVER POSSIBLE.
HOW ARE YOU, MARIANA DEAR?
HANGING IN THERE, PABLO...
JUSTINO'S DAUGHTER DIED.
UTERINE CANCER...
I HADN'T HEARD...
I'LL GO SEE HIM.

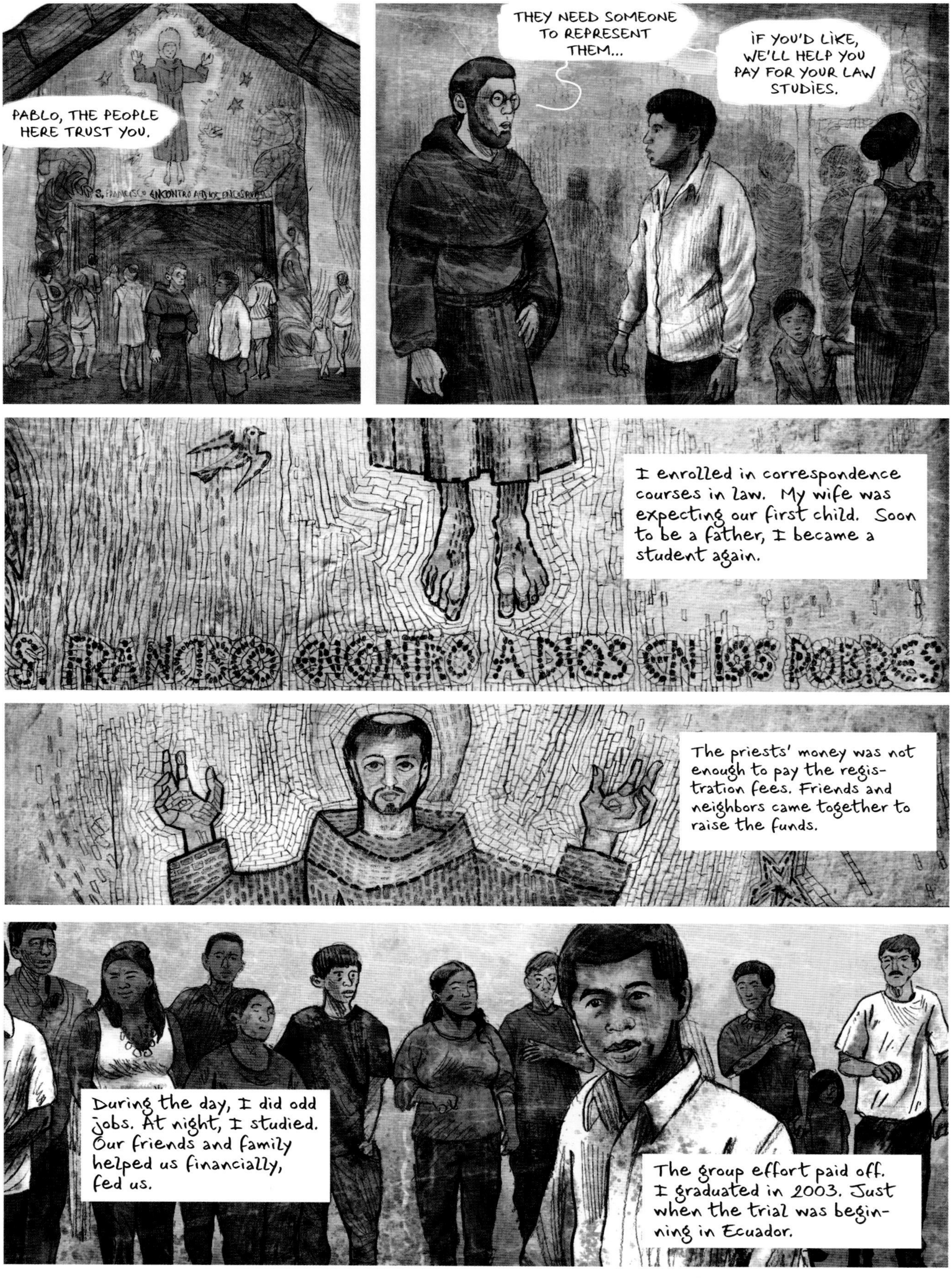
PABLO, THE PEOPLE HERE TRUST YOU.
THEY NEED SOMEONE TO REPRESENT THEM...
IF YOU'D LIKE, WE'LL HELP YOU PAY FOR YOUR LAW STUDIES.
I enrolled in correspondence courses in law. My wife was expecting our first child. Soon to be a father, I became a student again.
S. FRANCISCO ENCONTRO A DIOS EN LOS POBRES
The priests' money was not enough to pay the registration fees. Friends and neighbors came together to raise the funds.
During the day, I did odd jobs. At night, I studied. Our friends and family helped us financially, fed us.
The group effort paid off. I graduated in 2003. Just when the trial was beginning in Ecuador.

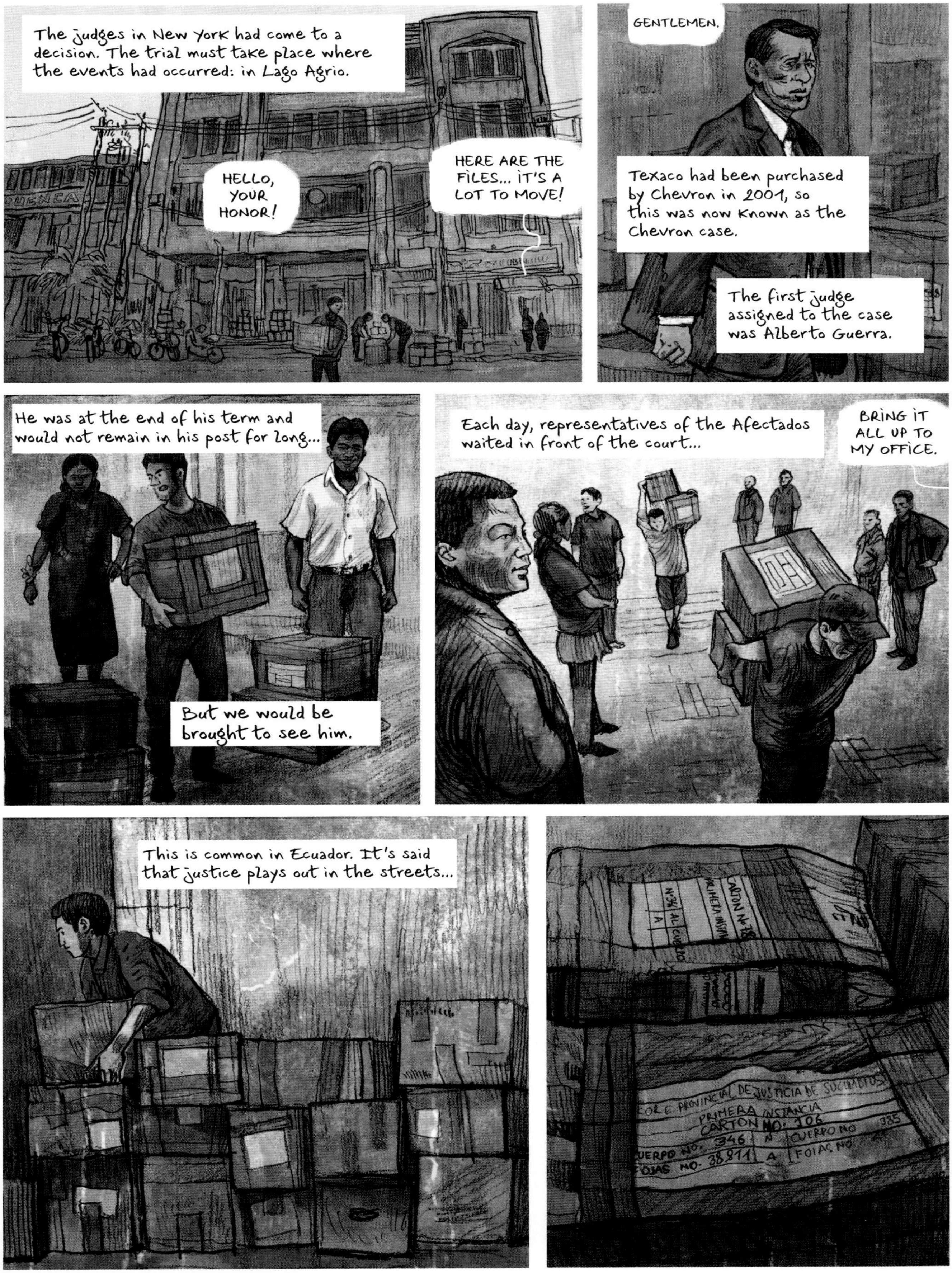
The judges in New York had come to a decision. The trial must take place where the events had occurred: in Lago Agrio.
HELLO, YOUR HONOR!
HERE ARE THE FILES... IT'S A LOT TO MOVE!
GENTLEMEN.
Texaco had been purchased by Chevron in 2001, so this was now known as the Chevron case.
The first judge assigned to the case was Alberto Guerra.
He was at the end of his term and would not remain in his post for long...
But we would be brought to see him.
Each day, representatives of the Afectados waited in front of the court...
BRING IT ALL UP TO MY OFFICE.
This is common in Ecuador. It's said that justice plays out in the streets...
PRIMERA INSTANCIA
CARTON NO. 106
CUERPO NO. 346
FOJAS NO. 38.811

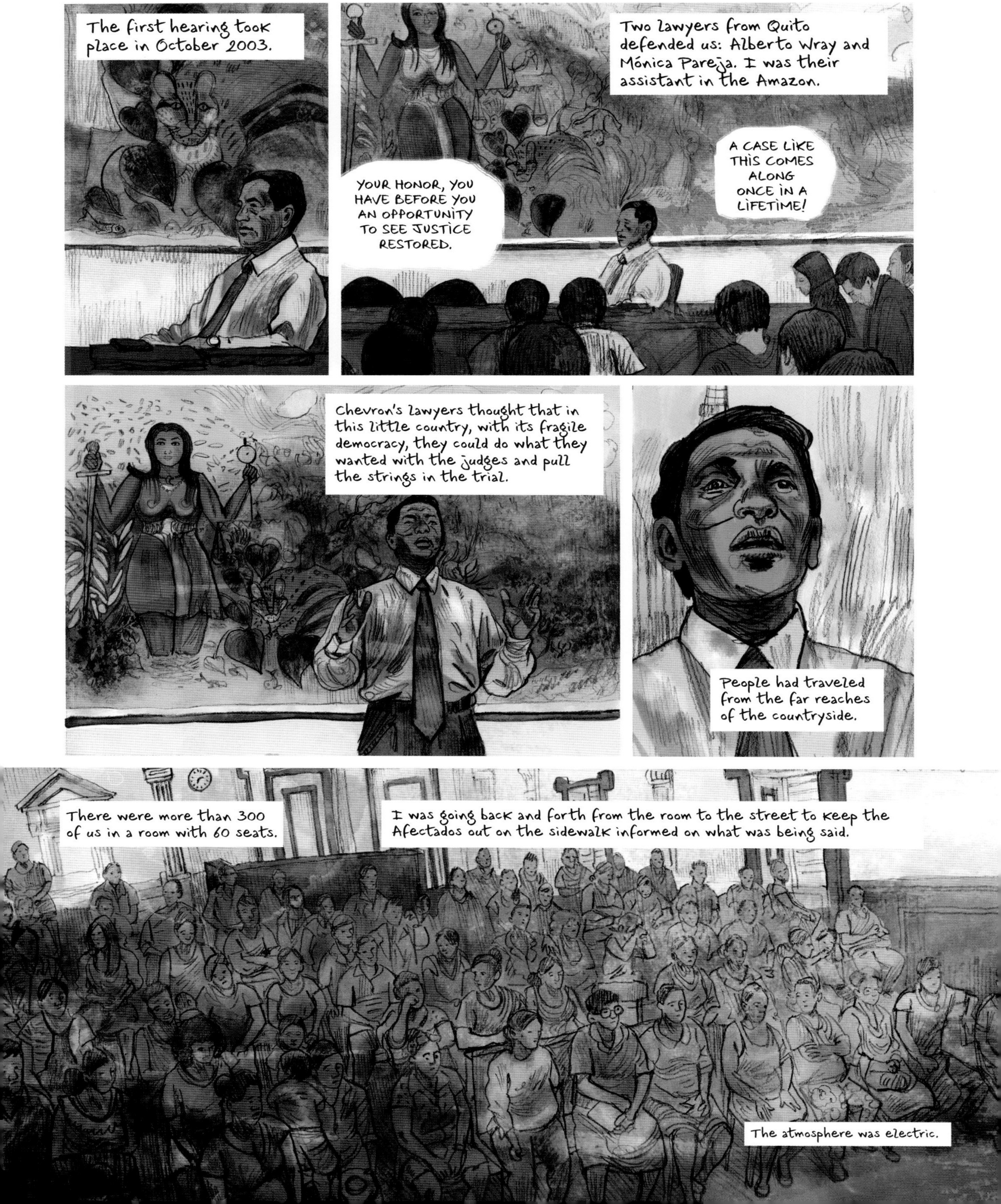
The first hearing took place in October 2003.
Two lawyers from Quito defended us: Alberto Wray and Mónica Pareja. I was their assistant in the Amazon.
YOUR HONOR, YOU HAVE BEFORE YOU AN OPPORTUNITY TO SEE JUSTICE RESTORED.
A CASE LIKE THIS COMES ALONG ONCE IN A LIFETIME!
Chevron's lawyers thought that in this little country, with its fragile democracy, they could do what they wanted with the judges and pull the strings in the trial.
People had traveled from the far reaches of the countryside.
There were more than 300 of us in a room with 60 seats.
I was going back and forth from the room to the street to keep the Afectados out on the sidewalk informed on what was being said.
The atmosphere was electric.

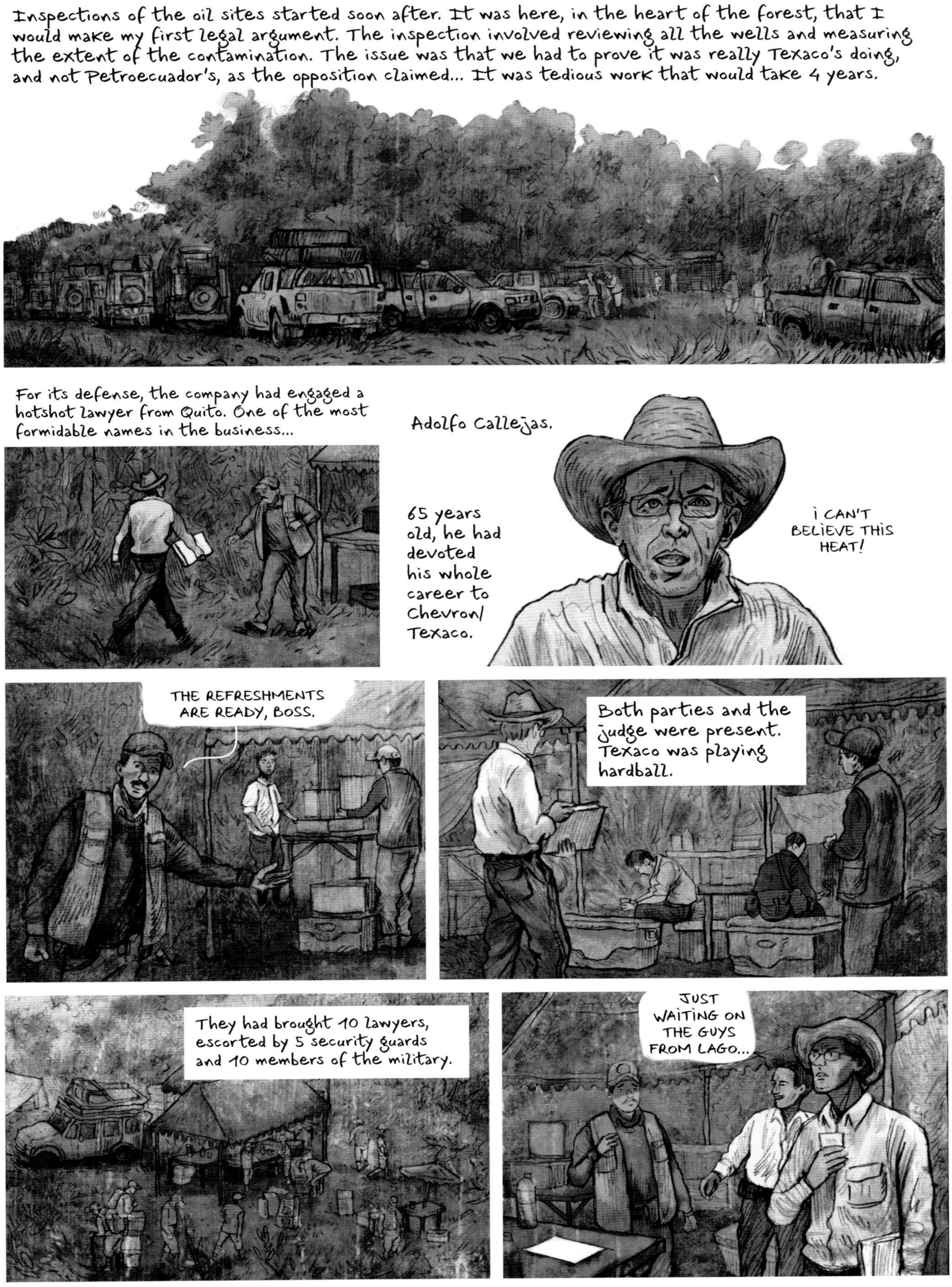
Inspections of the oil sites started soon after. It was here, in the heart of the forest, that I would make my first legal argument. The inspection involved reviewing all the wells and measuring the extent of the contamination. The issue was that we had to prove it was really Texaco's doing, and not Petroecuador's, as the opposition claimed... It was tedious work that would take 4 years.
For its defense, the company had engaged a hotshot lawyer from Quito. One of the most formidable names in the business...
Adolfo Callejas.
65 years old, he had devoted his whole career to Chevron/Texaco.
¡ CAN'T BELIEVE THIS HEAT!
THE REFRESHMENTS ARE READY, BOSS.
Both parties and the judge were present. Texaco was playing hardball.
They had brought 10 lawyers, escorted by 5 security guards and 10 members of the military.
JUST WAITING ON THE GUYS FROM LAGO...

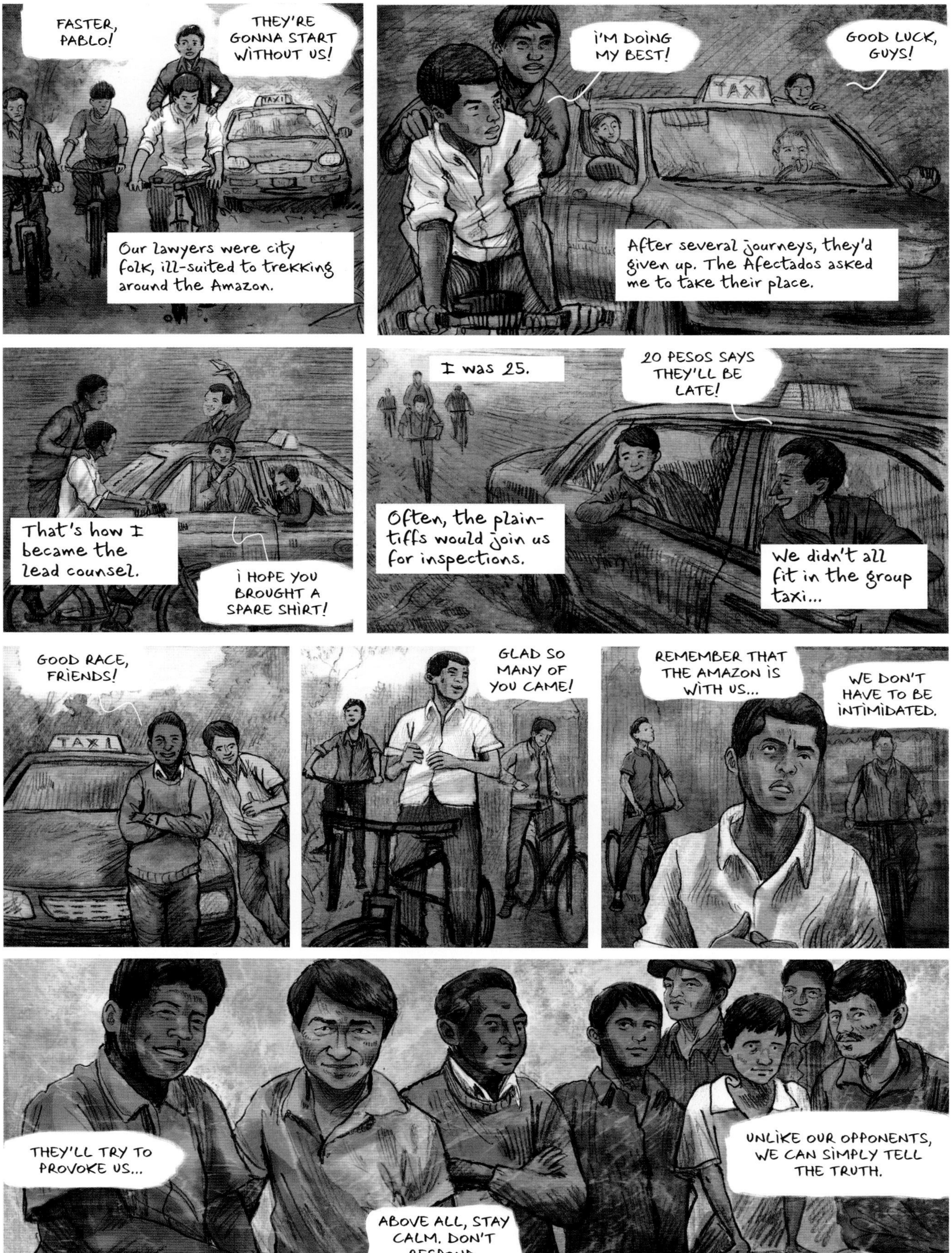
FASTER, PABLO!
THEY'RE GONNA START WITHOUT US!
Our lawyers were city folk, ill-suited to trekking around the Amazon.
I'M DOING MY BEST!
GOOD LUCK, GUYS!
After several journeys, they'd given up. The Afectados asked me to take their place.
That's how I became the lead counsel.
I HOPE YOU BROUGHT A SPARE SHIRT!
I was 25.
20 PESOS SAYS THEY'LL BE LATE!
Often, the plaintiffs would join us for inspections.
We didn't all fit in the group taxi...
GOOD RACE, FRIENDS!
GLAD SO MANY OF YOU CAME!
REMEMBER THAT THE AMAZON IS WITH US...
WE DON'T HAVE TO BE INTIMIDATED.
THEY'LL TRY TO PROVOKE US...
ABOVE ALL, STAY CALM. DON'T RESPOND.
UNLIKE OUR OPPONENTS, WE CAN SIMPLY TELL THE TRUTH.

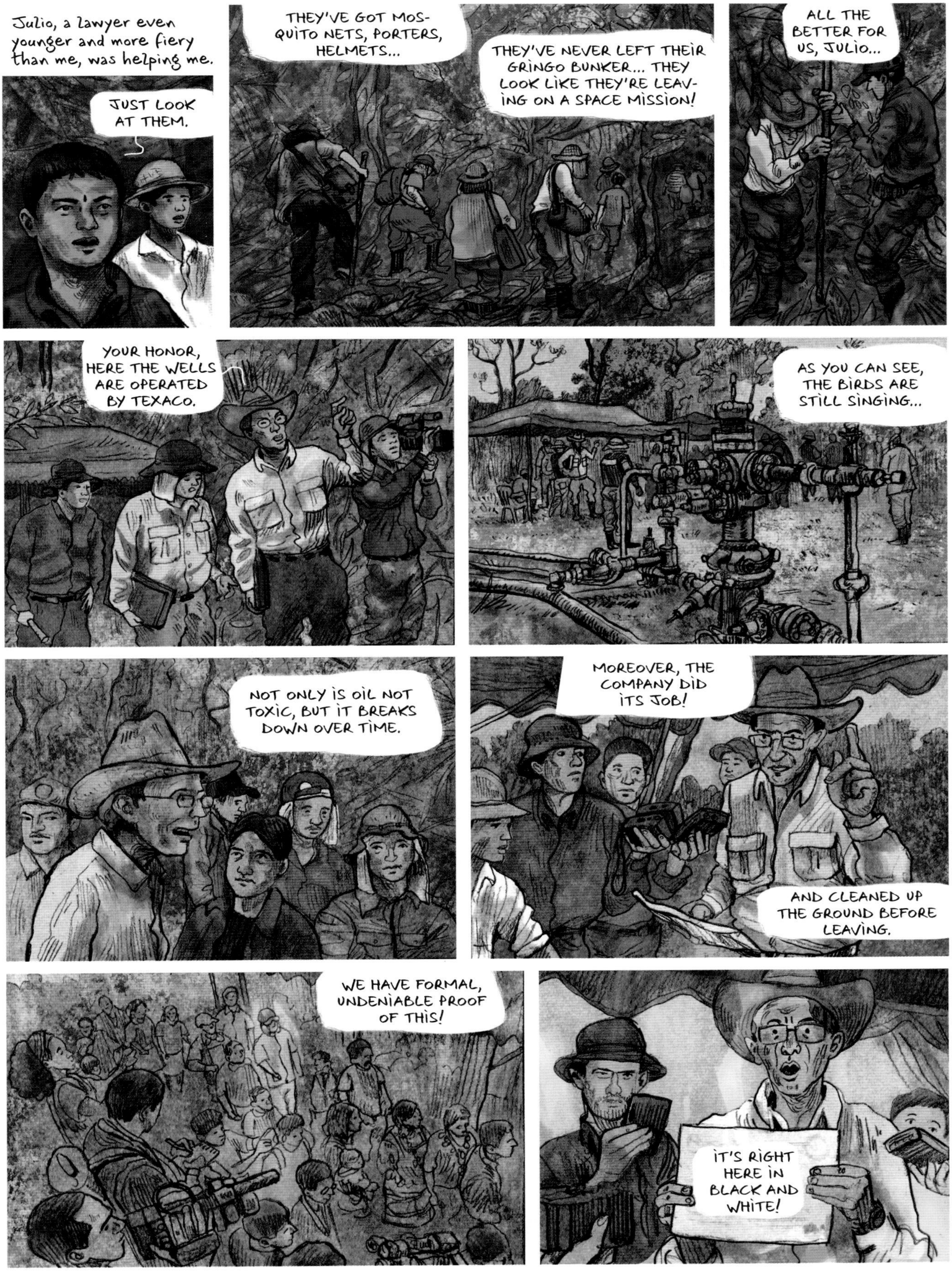
Julio, a lawyer even younger and more fiery than me, was helping me.
JUST LOOK AT THEM.
THEY'VE GOT MOSQUITO NETS, PORTERS, HELMETS...
THEY'VE NEVER LEFT THEIR GRINGO BUNKER... THEY LOOK LIKE THEY'RE LEAVING ON A SPACE MISSION!
ALL THE BETTER FOR US, JULIO...
YOUR HONOR, HERE THE WELLS ARE OPERATED BY TEXACO.
AS YOU CAN SEE, THE BIRDS ARE STILL SINGING...
NOT ONLY IS OIL NOT TOXIC, BUT IT BREAKS DOWN OVER TIME.
MOREOVER, THE COMPANY DID ITS JOB!
AND CLEANED UP THE GROUND BEFORE LEAVING.
WE HAVE FORMAL, UNDENIABLE PROOF OF THIS!
IT'S RIGHT HERE IN BLACK AND WHITE!

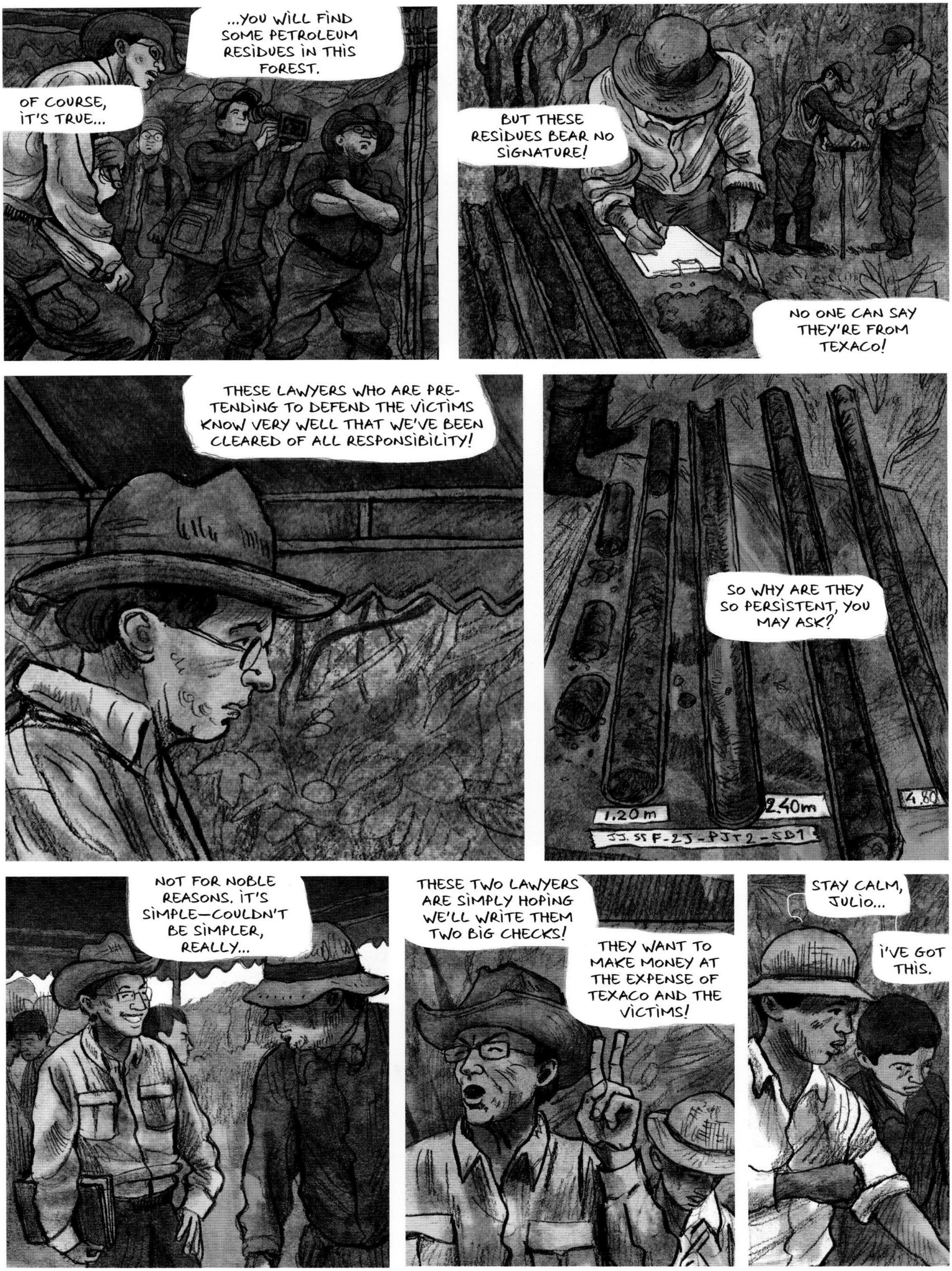
...YOU WILL FIND SOME PETROLEUM RESIDUES IN THIS FOREST.
OF COURSE, IT'S TRUE...
BUT THESE RESIDUES BEAR NO SIGNATURE!
NO ONE CAN SAY THEY'RE FROM TEXACO!
THESE LAWYERS WHO ARE PRETENDING TO DEFEND THE VICTIMS KNOW VERY WELL THAT WE'VE BEEN CLEARED OF ALL RESPONSIBILITY!
SO WHY ARE THEY SO PERSISTENT, YOU MAY ASK?
1.20m
2.40m
4.80
NOT FOR NOBLE REASONS. IT'S SIMPLE—COULDN'T BE SIMPLER, REALLY...
THESE TWO LAWYERS ARE SIMPLY HOPING WE'LL WRITE THEM TWO BIG CHECKS!
THEY WANT TO MAKE MONEY AT THE EXPENSE OF TEXACO AND THE VICTIMS!
STAY CALM, JULIO...
I'VE GOT THIS.

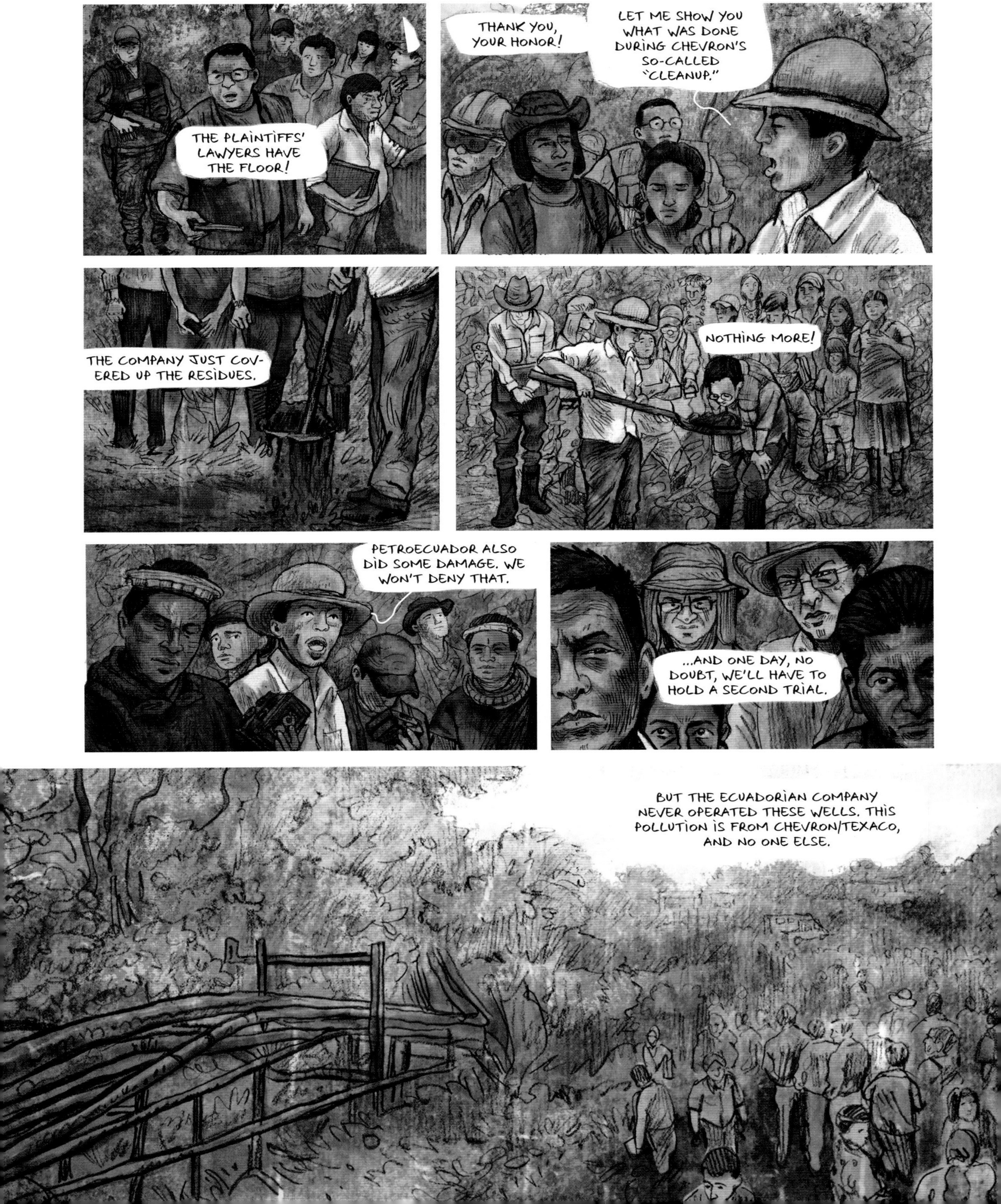
THE PLAINTIFFS' LAWYERS HAVE THE FLOOR!
THANK YOU, YOUR HONOR!
LET ME SHOW YOU WHAT WAS DONE DURING CHEVRON'S SO-CALLED "CLEANUP."
THE COMPANY JUST COVERED UP THE RESIDUES.
NOTHING MORE!
PETROECUADOR ALSO DID SOME DAMAGE. WE WON'T DENY THAT.
...AND ONE DAY, NO DOUBT, WE'LL HAVE TO HOLD A SECOND TRIAL.
BUT THE ECUADORIAN COMPANY NEVER OPERATED THESE WELLS. THIS POLLUTION IS FROM CHEVRON/TEXACO, AND NO ONE ELSE.

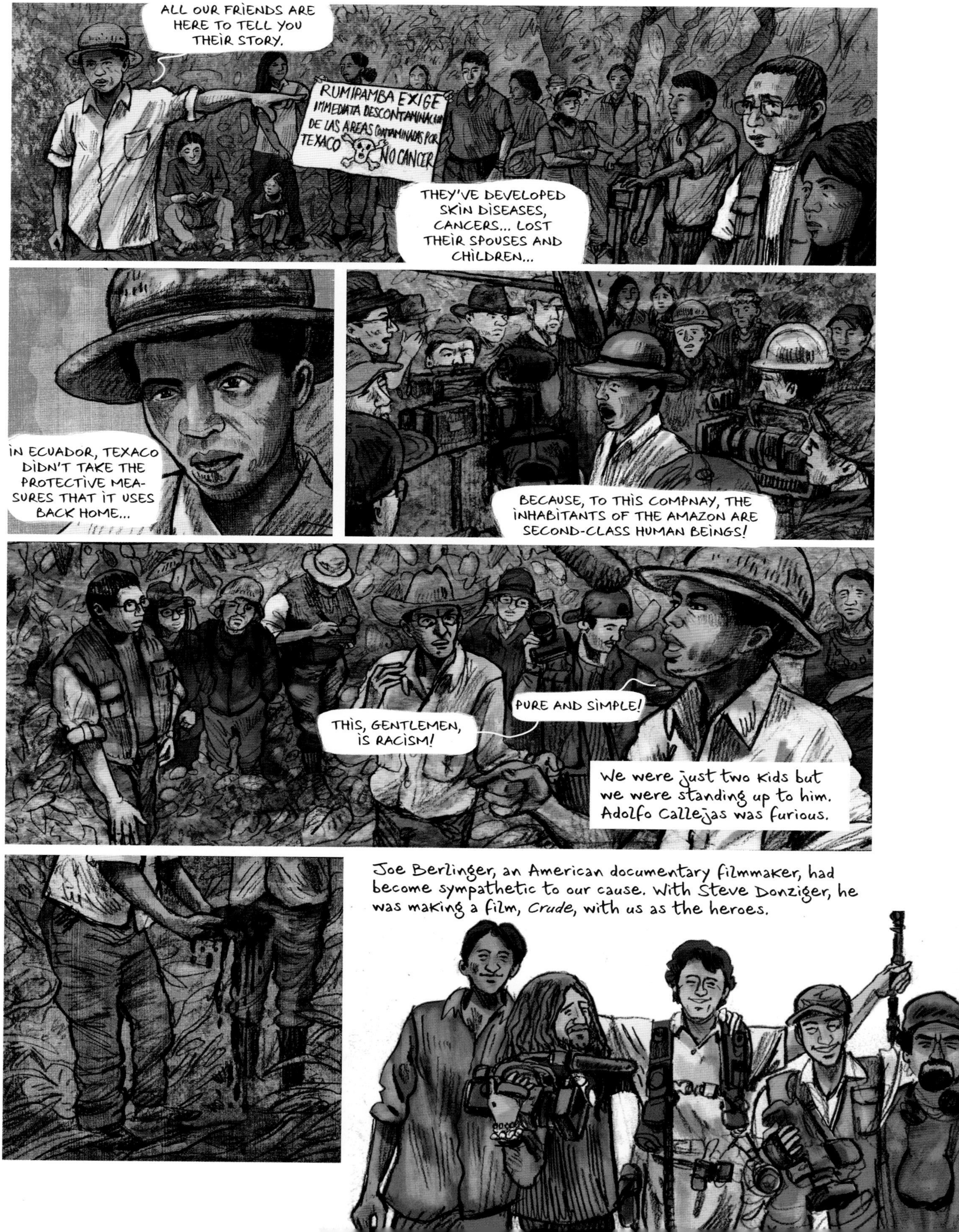
ALL OUR FRIENDS ARE HERE TO TELL YOU THEIR STORY.
RUMIPAMBA EXIGE
IMMEDIATA DESCONTAMINAC
DE LAS AREAS CONTAMINADAS POR
TEXACO
NO CANCER
THEY'VE DEVELOPED SKIN DISEASES, CANCERS... LOST THEIR SPOUSES AND CHILDREN...
IN ECUADOR, TEXACO DIDN'T TAKE THE PROTECTIVE MEASURES THAT IT USES BACK HOME...
BECAUSE, TO THIS COMPNAY, THE INHABITANTS OF THE AMAZON ARE SECOND-CLASS HUMAN BEINGS!
THIS, GENTLEMEN, IS RACISM!
PURE AND SIMPLE!
We were just two kids but we were standing up to him. Adolfo Callejas was furious.
Joe Berlinger, an American documentary filmmaker, had become sympathetic to our cause. With Steve Donziger, he was making a film, *Crude*, with us as the heroes.

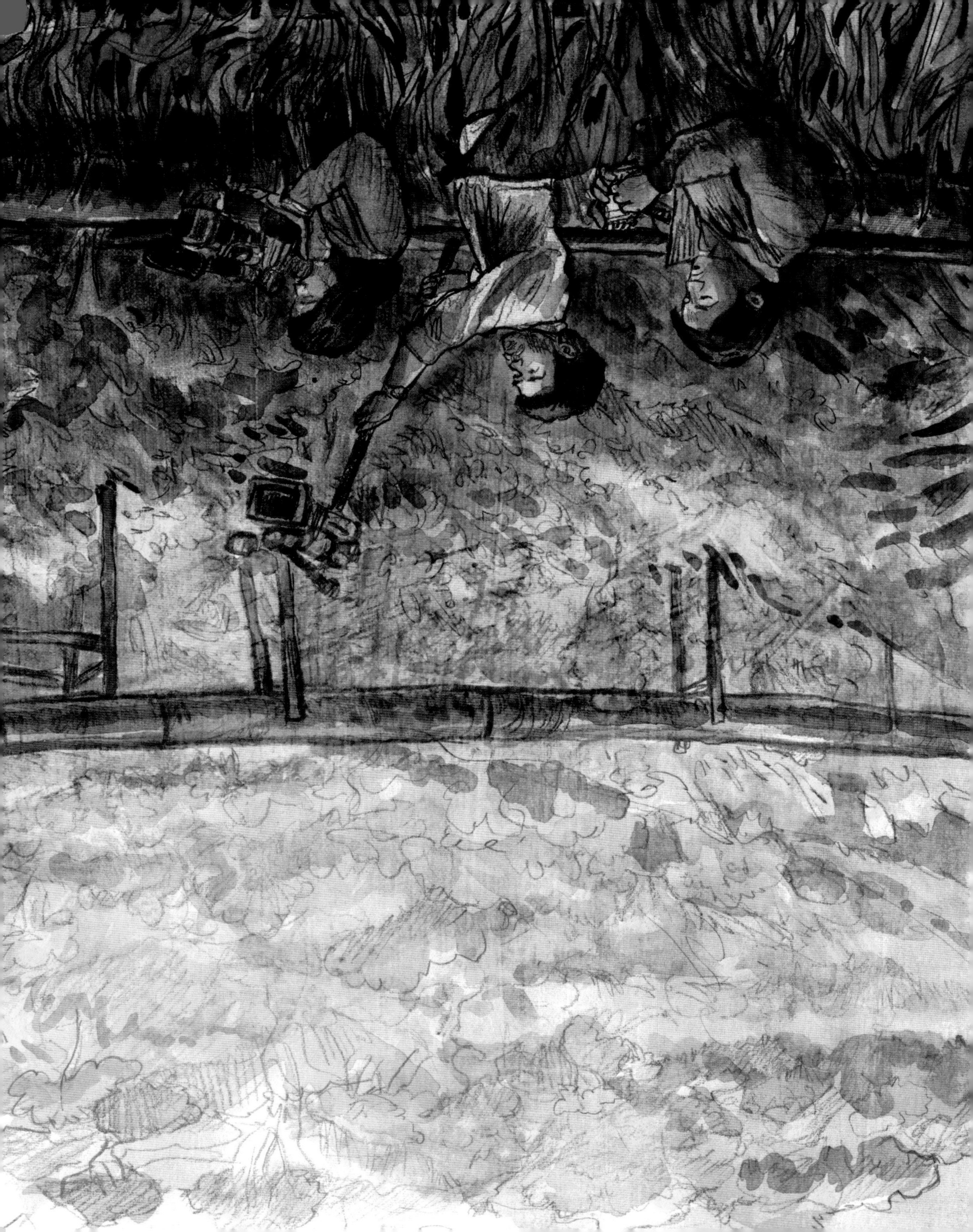

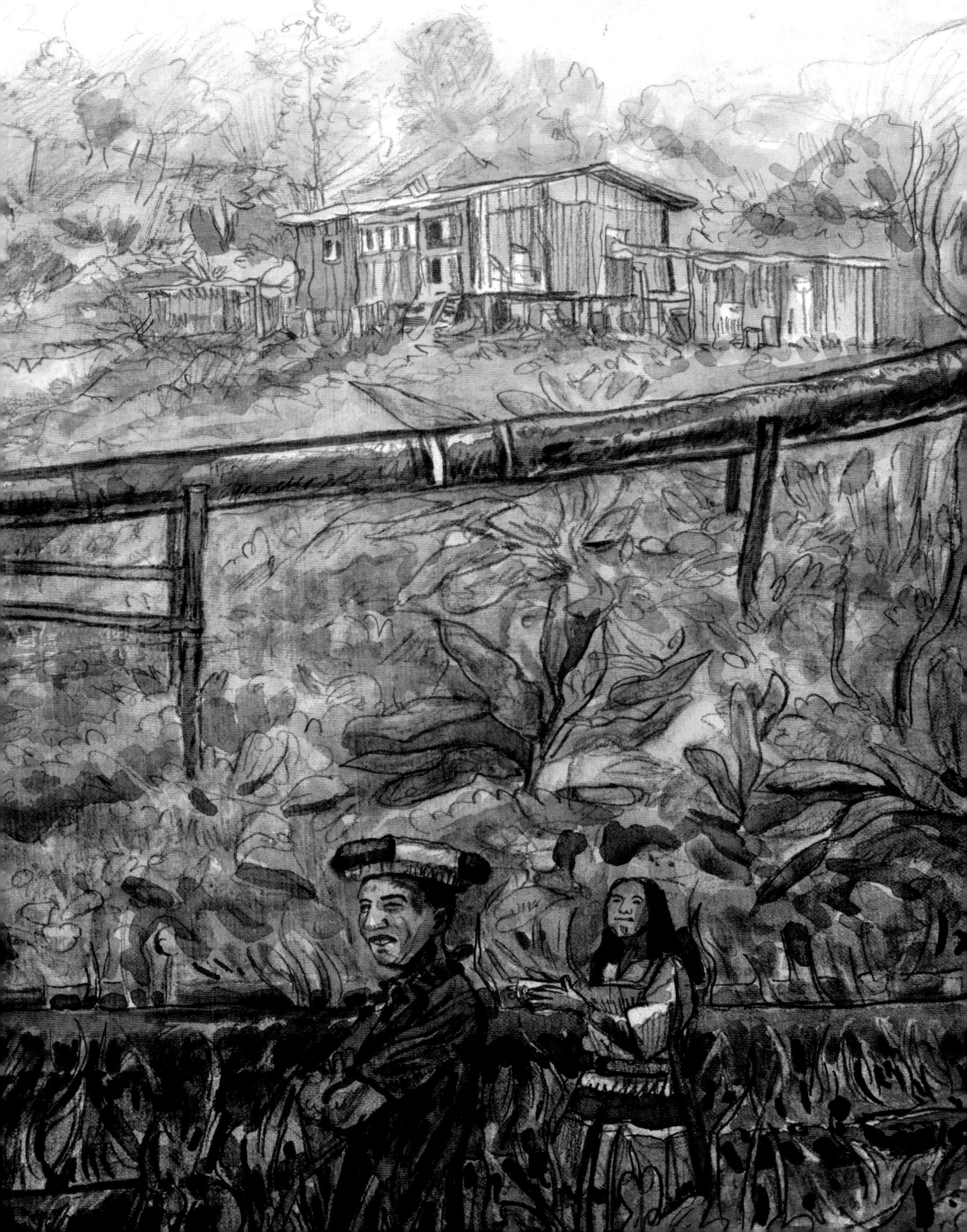

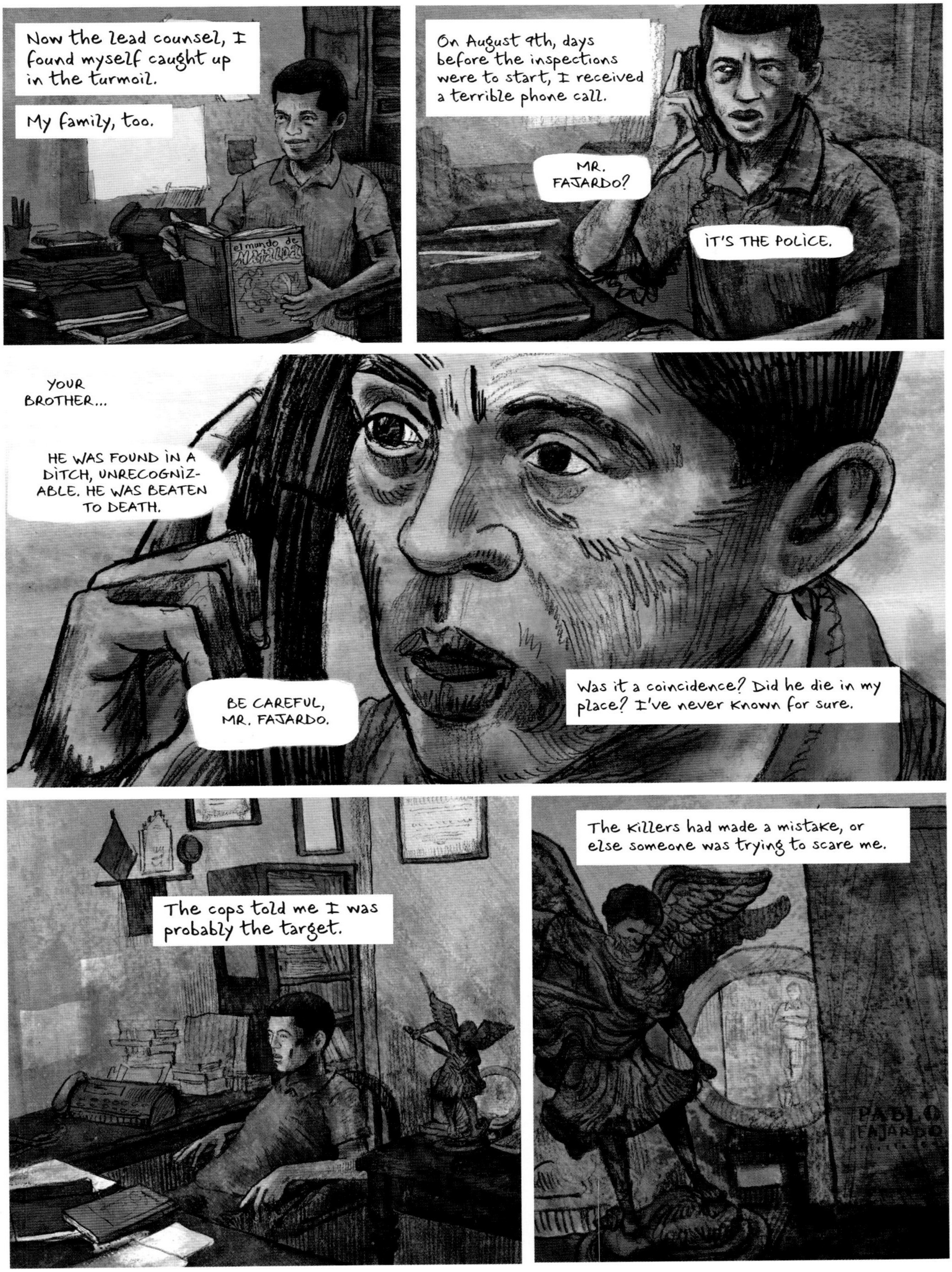
Now the lead counsel, I found myself caught up in the turmoil.
My family, too.
On August 9th, days before the inspections were to start, I received a terrible phone call.
MR. FAJARDO?
IT'S THE POLICE.
YOUR BROTHER...
HE WAS FOUND IN A DITCH, UNRECOGNIZABLE. HE WAS BEATEN TO DEATH.
BE CAREFUL, MR. FAJARDO.
Was it a coincidence? Did he die in my place? I've never known for sure.
The cops told me I was probably the target.
The killers had made a mistake, or else someone was trying to scare me.
PABLO FAJARDO

The police advised me to carry a weapon. I never could bring myself to.

Whenever I said goodbye to a friend, a loved one, I couldn't shake the feeling that it might be the last time I saw them.

But the hardest thing was the guilt. My mother and brothers believed Wilson was dead because of me.

Spies started hanging around my house. I was afraid for my children.

My daughter and my son, who had just been born, went to live with their maternal grandparents, 12 miles from Lago Agrio.

That's where they grew up. They never returned to live in our house.

On weekends, I'd go and spend a day with them.

Sus Padres y hermanos

Padre Jose Fajardo

Madre Marin Mendoza

Wilson Oriol Fajardo Mendoza

★ Sep. 22 - 1975

✝ Agosto

I tried as best I could to maintain a normal life.

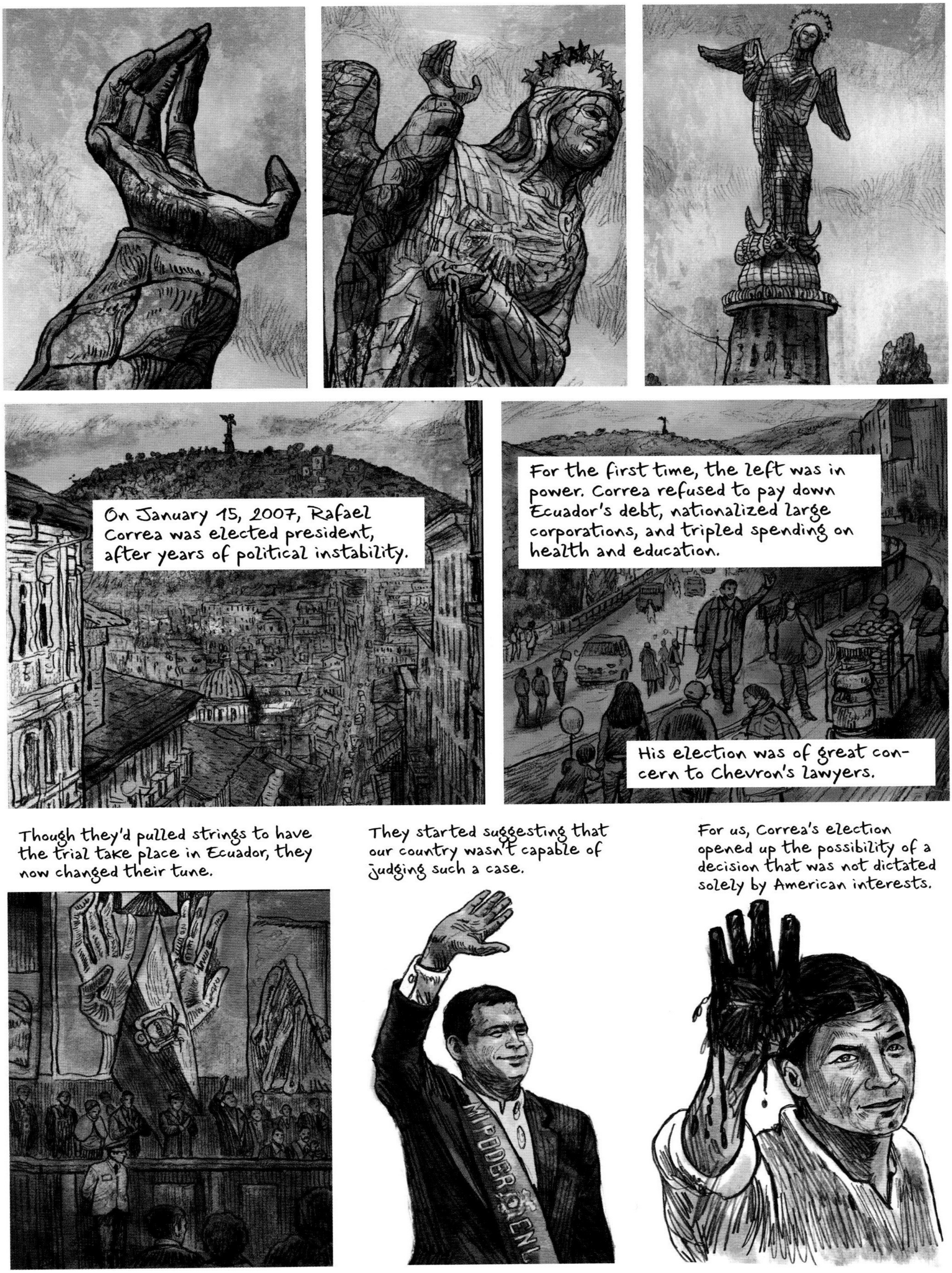
On January 15, 2007, Rafael Correa was elected president, after years of political instability.
For the first time, the left was in power. Correa refused to pay down Ecuador's debt, nationalized large corporations, and tripled spending on health and education.
His election was of great concern to Chevron's lawyers.
Though they'd pulled strings to have the trial take place in Ecuador, they now changed their tune.
They started suggesting that our country wasn't capable of judging such a case.
MI PODER EN L
For us, Correa's election opened up the possibility of a decision that was not dictated solely by American interests.

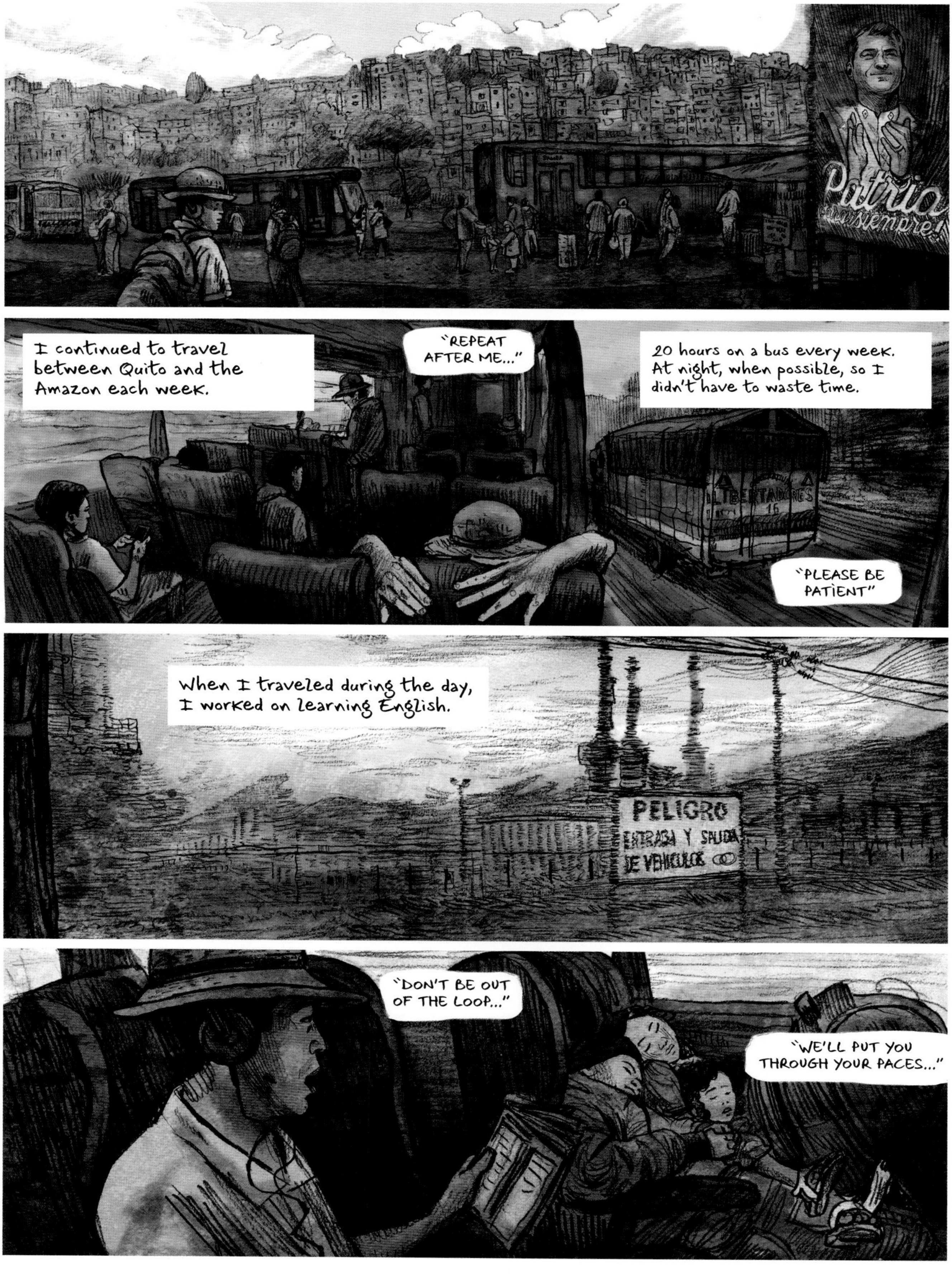
Patria
I continued to travel between Quito and the Amazon each week.
"REPEAT AFTER ME..."
20 hours on a bus every week. At night, when possible, so I didn't have to waste time.
LIBERTADORES
"PLEASE BE PATIENT"
When I traveled during the day, I worked on learning English.
PELIGRO
ENTRADA Y SALIDA
DE VEHICULOS
"DON'T BE OUT OF THE LOOP..."
"WE'LL PUT YOU THROUGH YOUR PACES..."

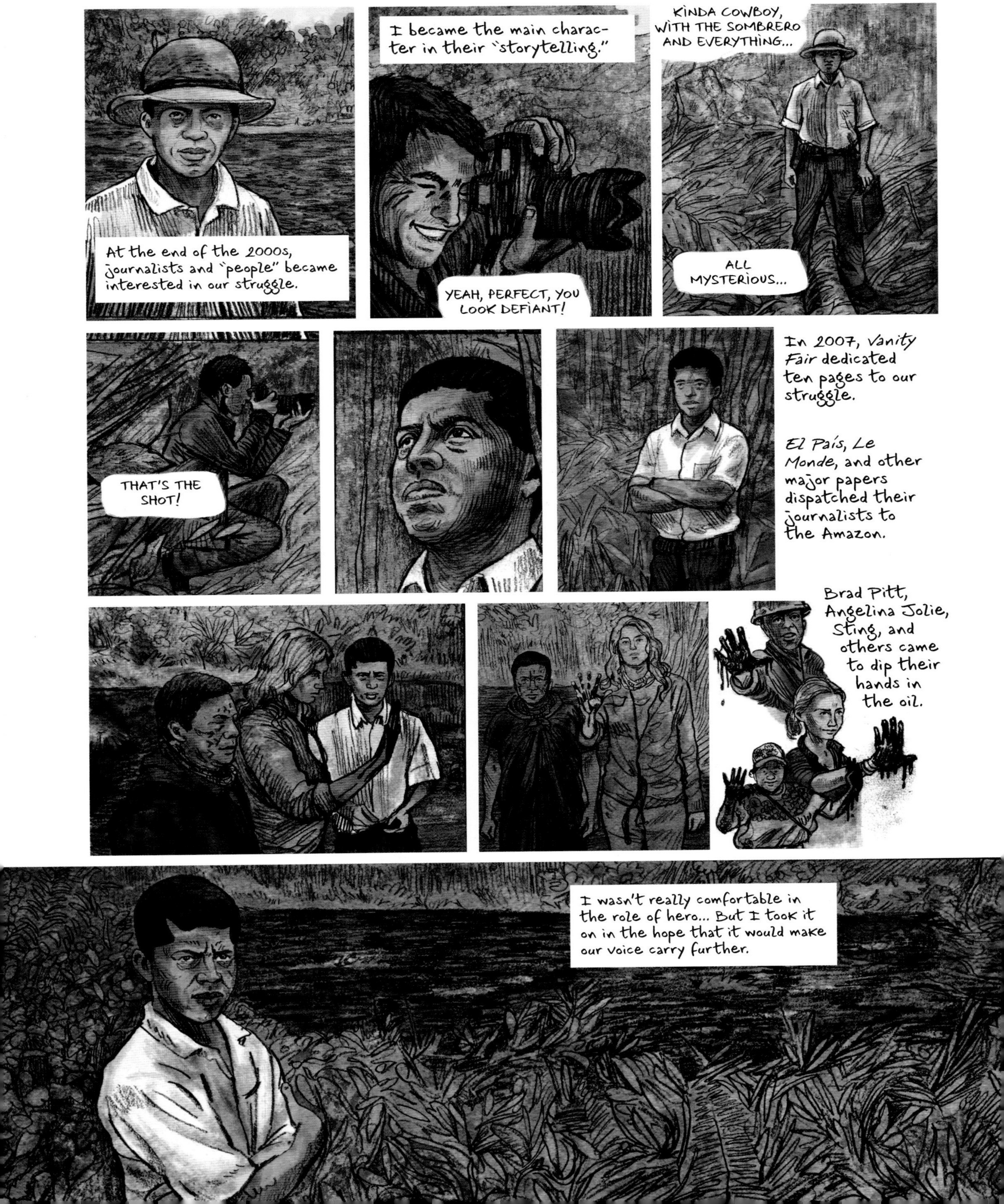
At the end of the 2000s, journalists and "people" became interested in our struggle.
I became the main character in their "storytelling."
YEAH, PERFECT, YOU LOOK DEFIANT!
KINDA COWBOY, WITH THE SOMBRERO AND EVERYTHING...
ALL MYSTERIOUS...
THAT'S THE SHOT!
In 2007, *Vanity Fair* dedicated ten pages to our struggle.
El País, *Le Monde*, and other major papers dispatched their journalists to the Amazon.
Brad Pitt, Angelina Jolie, Sting, and others came to dip their hands in the oil.
I wasn't really comfortable in the role of hero... But I took it on in the hope that it would make our voice carry further.

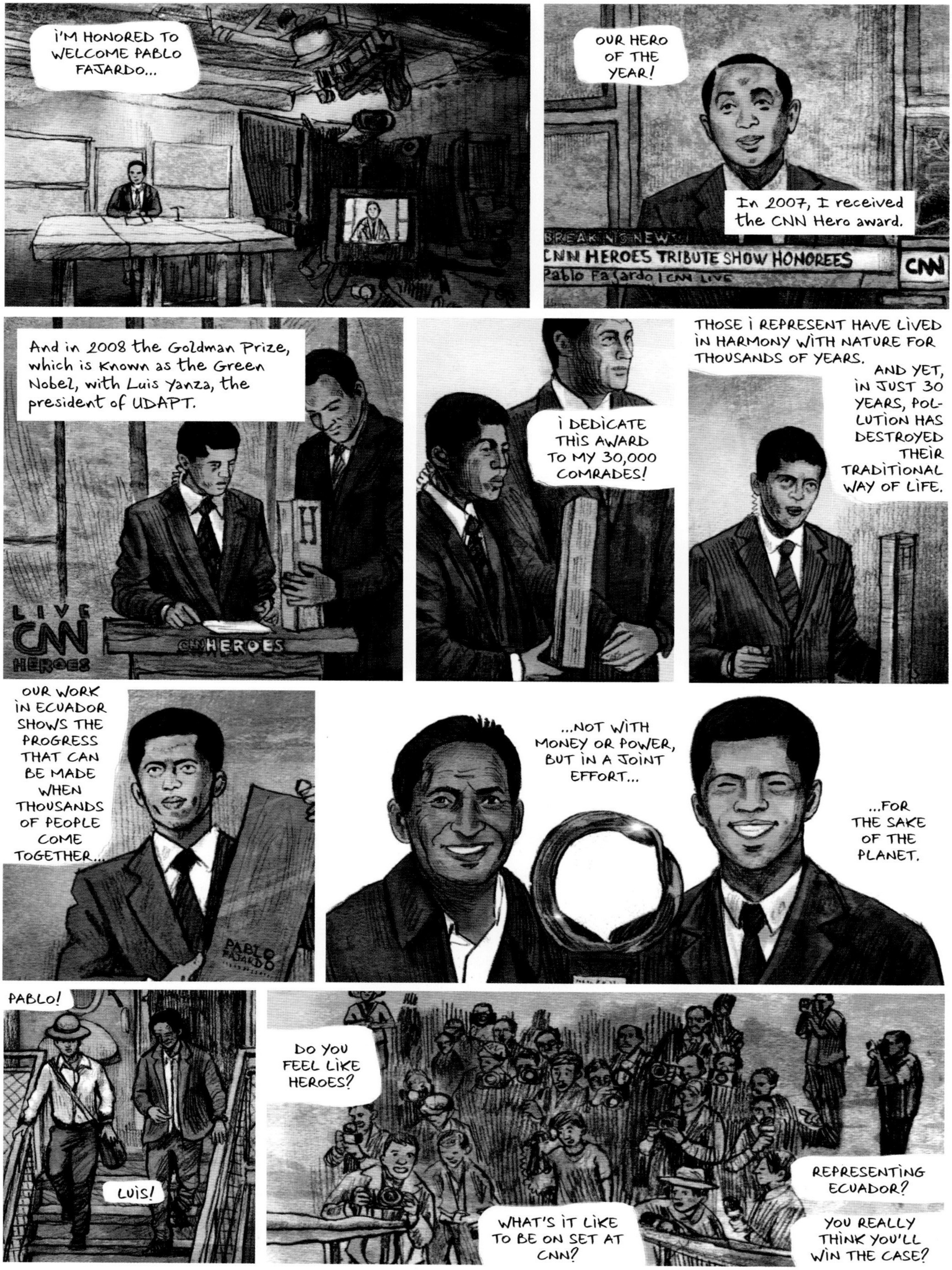
I'M HONORED TO WELCOME PABLO FAJARDO...
OUR HERO OF THE YEAR!
In 2007, I received the CNN Hero award.
BREAKING NEWS
CNN HEROES TRIBUTE SHOW HONOREES
Pablo Fajardo | CNN LIVE
CNN
And in 2008 the Goldman Prize, which is known as the Green Nobel, with Luis Yanza, the president of UDAPT.
LIVE
CNN
HEROES
CNN HEROES
I DEDICATE THIS AWARD TO MY 30,000 COMRADES!
THOSE I REPRESENT HAVE LIVED IN HARMONY WITH NATURE FOR THOUSANDS OF YEARS.
AND YET, IN JUST 30 YEARS, POLLUTION HAS DESTROYED THEIR TRADITIONAL WAY OF LIFE.
OUR WORK IN ECUADOR SHOWS THE PROGRESS THAT CAN BE MADE WHEN THOUSANDS OF PEOPLE COME TOGETHER...
PABLO FAJARDO
...NOT WITH MONEY OR POWER, BUT IN A JOINT EFFORT...
...FOR THE SAKE OF THE PLANET.
PABLO!
LUIS!
DO YOU FEEL LIKE HEROES?
REPRESENTING ECUADOR?
WHAT'S IT LIKE TO BE ON SET AT CNN?
YOU REALLY THINK YOU'LL WIN THE CASE?

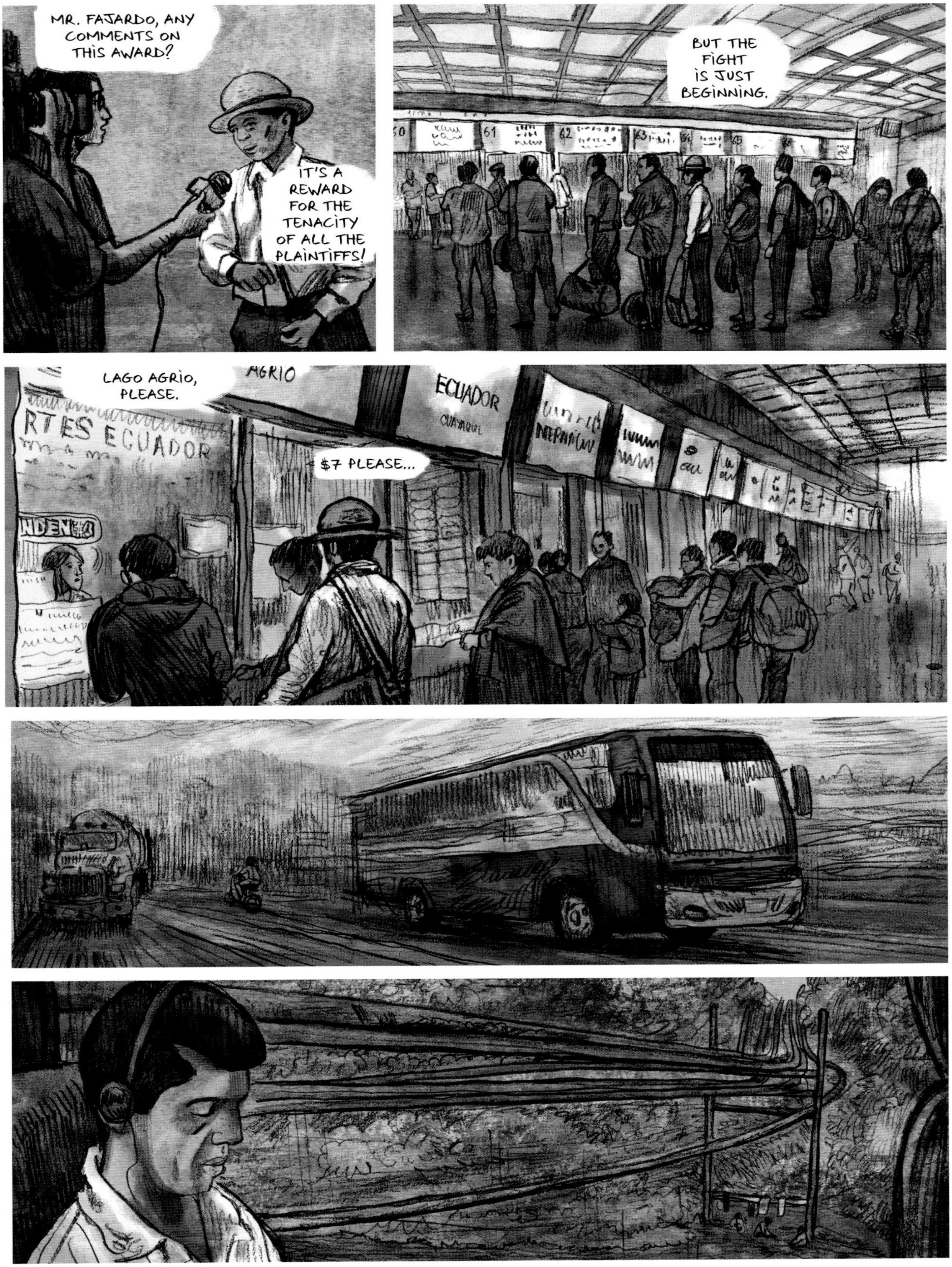
MR. FAJARDO, ANY COMMENTS ON THIS AWARD?
IT'S A REWARD FOR THE TENACITY OF ALL THE PLAINTIFFS!
BUT THE FIGHT IS JUST BEGINNING.
LAGO AGRIO, PLEASE.
AGRIO
ECUADOR
RTES ECUADOR
$7 PLEASE...

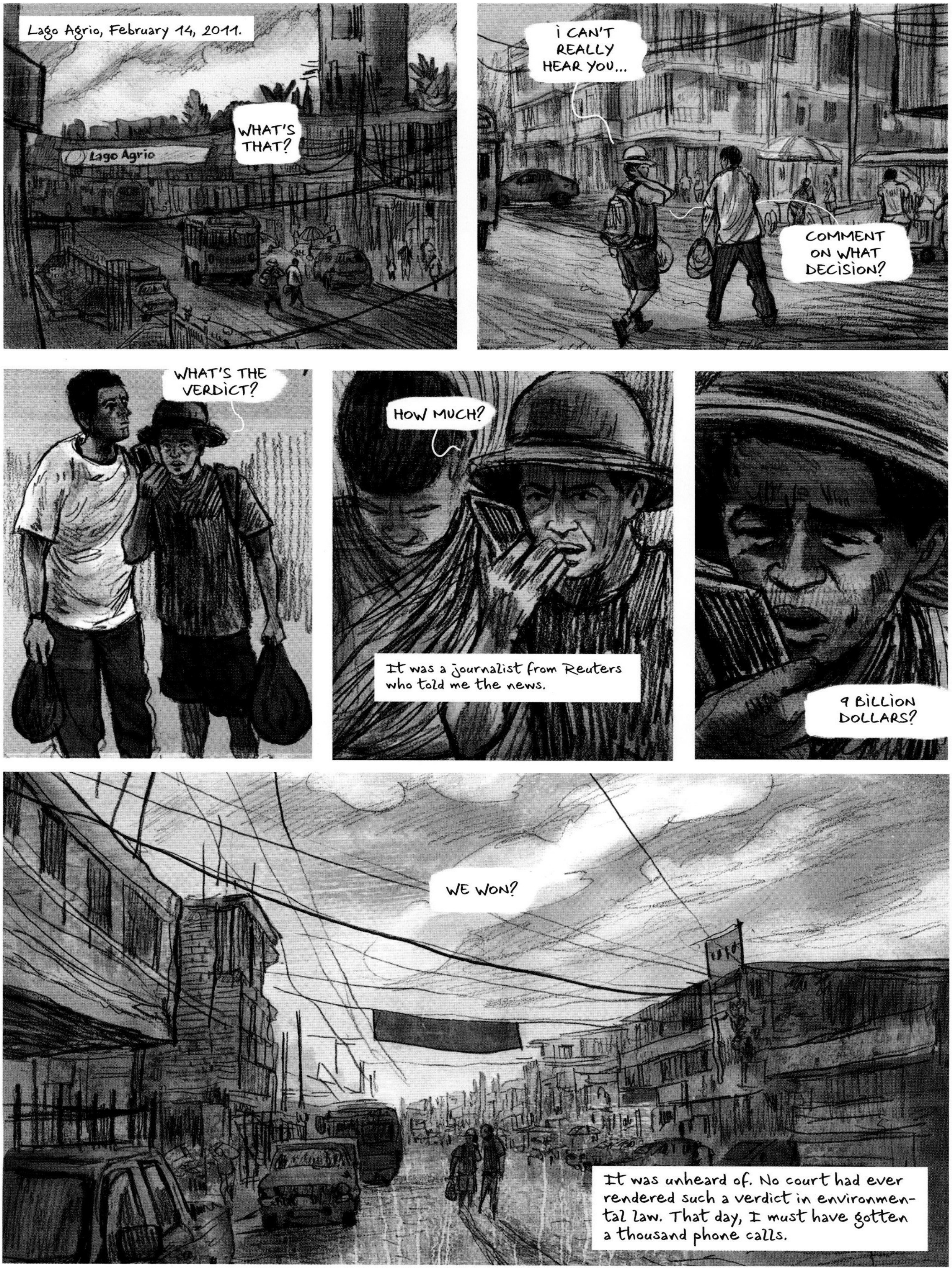
Lago Agrio, February 14, 2011.
Lago Agrio
WHAT'S THAT?
I CAN'T REALLY HEAR YOU...
COMMENT ON WHAT DECISION?
WHAT'S THE VERDICT?
HOW MUCH?
It was a journalist from Reuters who told me the news.
9 BILLION DOLLARS?
WE WON?
It was unheard of. No court had ever rendered such a verdict in environmental law. That day, I must have gotten a thousand phone calls.

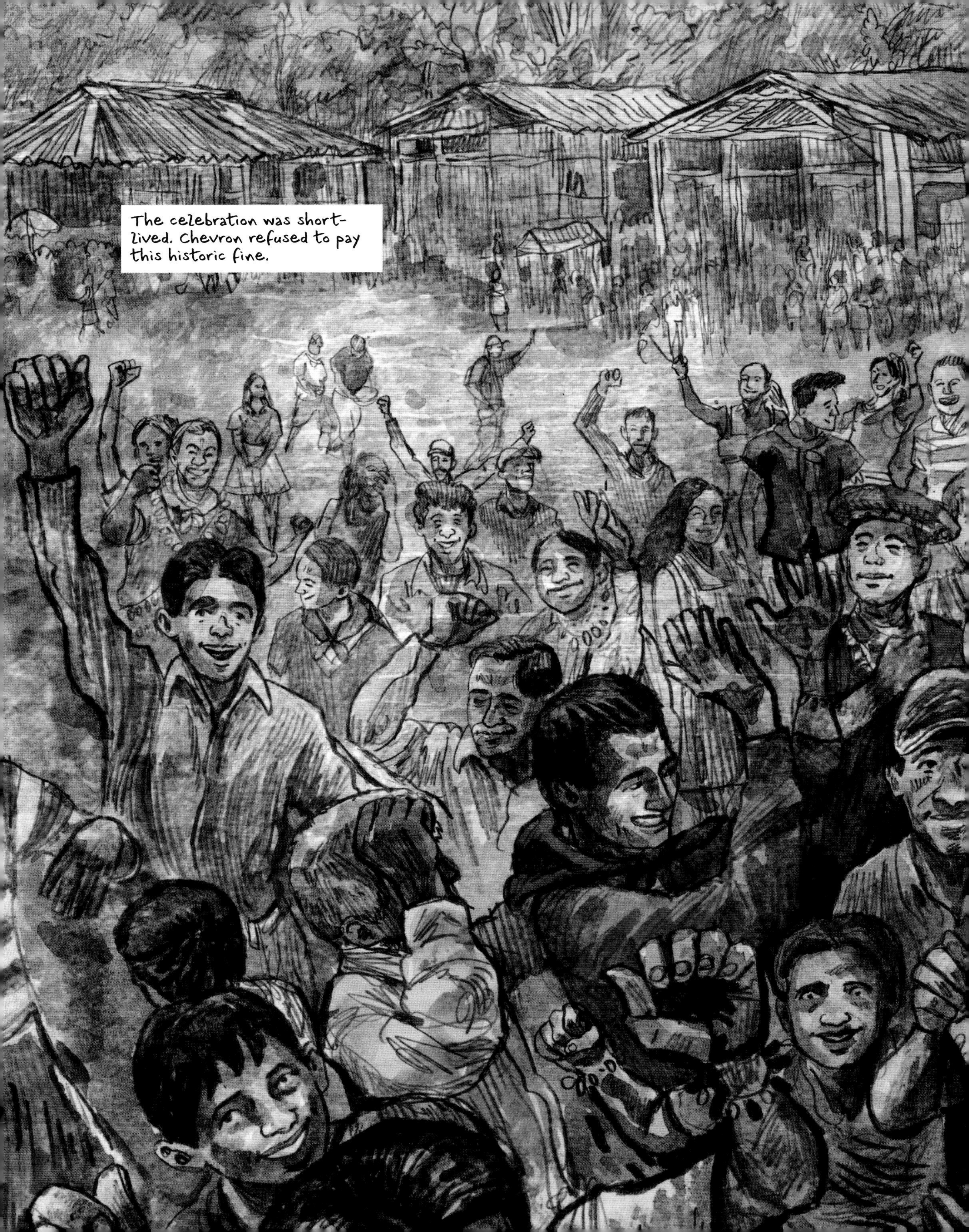
The celebration was short-lived. Chevron refused to pay this historic fine.

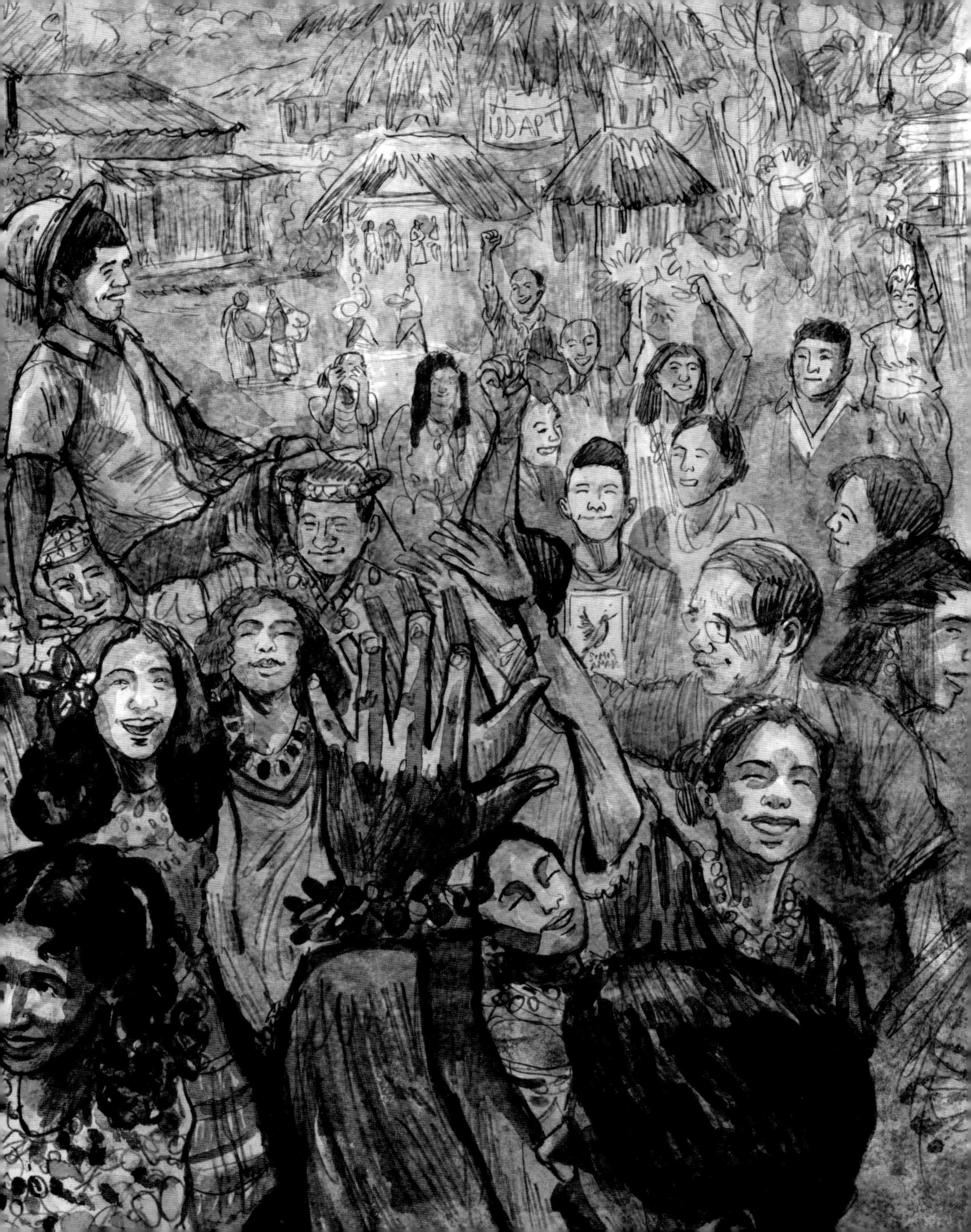
UDAPT

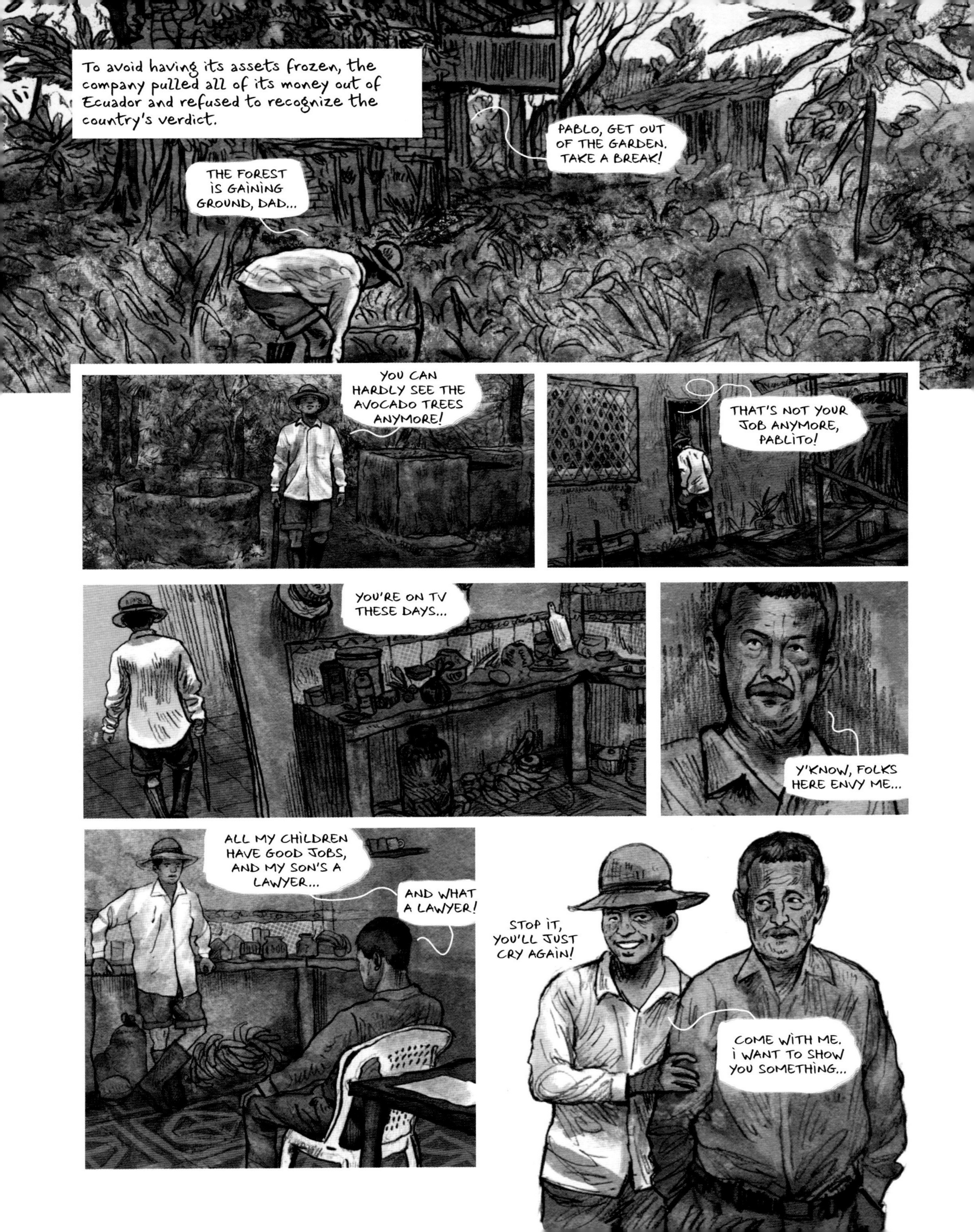
To avoid having its assets frozen, the company pulled all of its money out of Ecuador and refused to recognize the country's verdict.
PABLO, GET OUT OF THE GARDEN. TAKE A BREAK!
THE FOREST IS GAINING GROUND, DAD...
YOU CAN HARDLY SEE THE AVOCADO TREES ANYMORE!
THAT'S NOT YOUR JOB ANYMORE, PABLITO!
YOU'RE ON TV THESE DAYS...
Y'KNOW, FOLKS HERE ENVY ME...
ALL MY CHILDREN HAVE GOOD JOBS, AND MY SON'S A LAWYER...
AND WHAT A LAWYER!
STOP IT, YOU'LL JUST CRY AGAIN!
COME WITH ME. I WANT TO SHOW YOU SOMETHING...

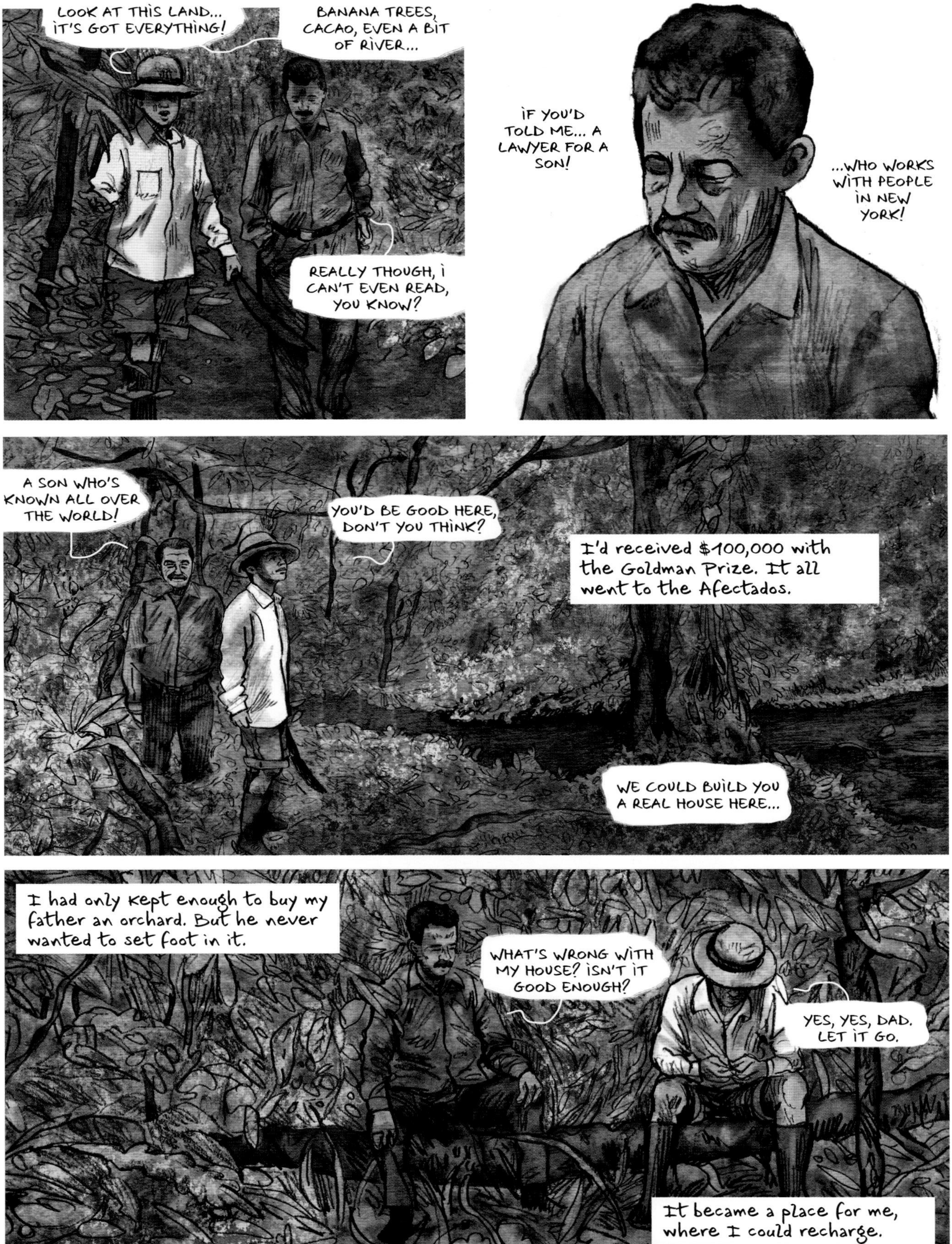
LOOK AT THIS LAND... IT'S GOT EVERYTHING!
BANANA TREES, CACAO, EVEN A BIT OF RIVER...
REALLY THOUGH, I CAN'T EVEN READ, YOU KNOW?
IF YOU'D TOLD ME... A LAWYER FOR A SON!
...WHO WORKS WITH PEOPLE IN NEW YORK!
A SON WHO'S KNOWN ALL OVER THE WORLD!
YOU'D BE GOOD HERE, DON'T YOU THINK?
I'd received $100,000 with the Goldman Prize. It all went to the Afectados.
WE COULD BUILD YOU A REAL HOUSE HERE...
I had only kept enough to buy my father an orchard. But he never wanted to set foot in it.
WHAT'S WRONG WITH MY HOUSE? ISN'T IT GOOD ENOUGH?
YES, YES, DAD. LET IT GO.
It became a place for me, where I could recharge.

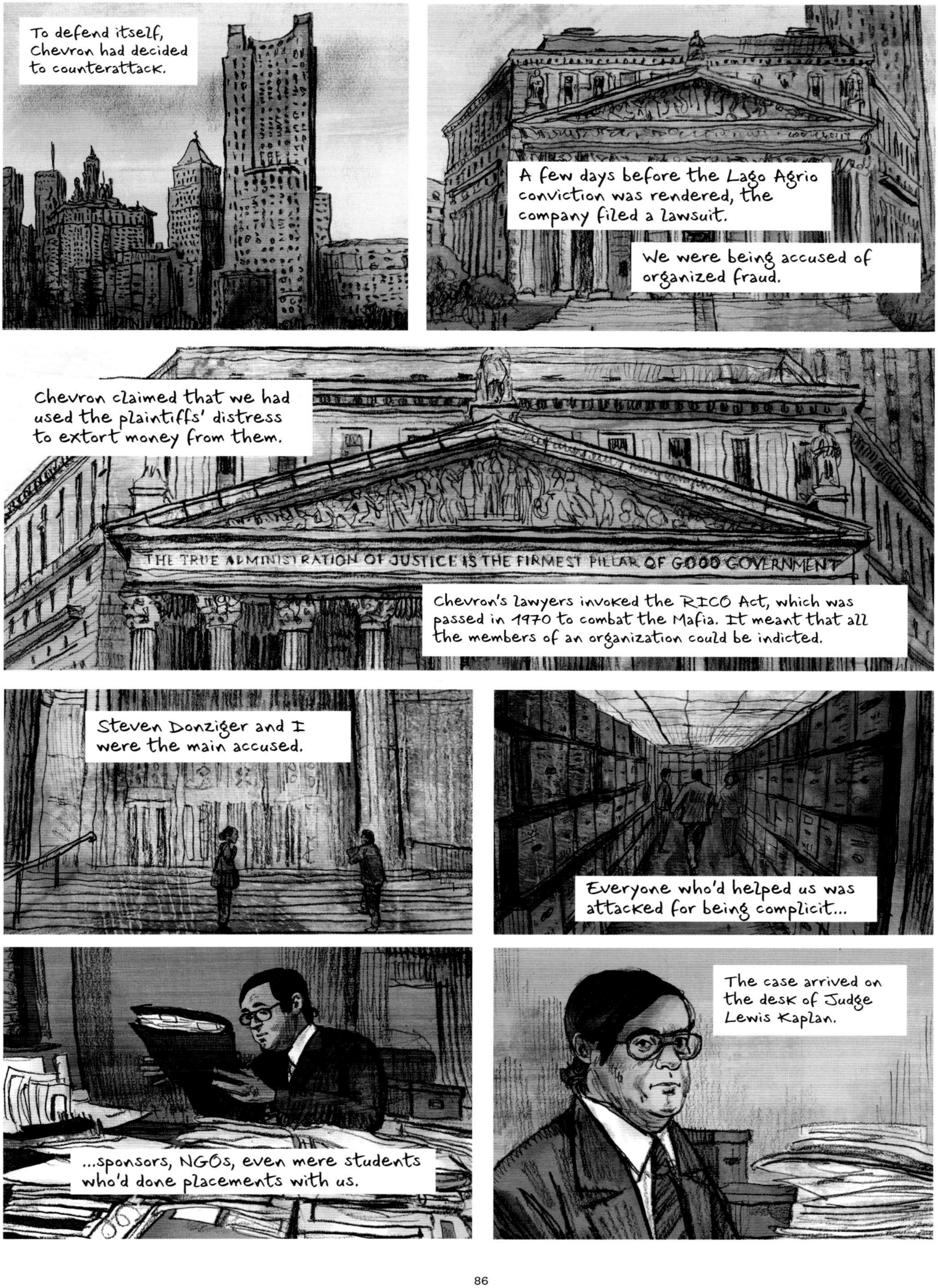
To defend itself, Chevron had decided to counterattack.
A few days before the Lago Agrio conviction was rendered, the company filed a lawsuit.
We were being accused of organized fraud.
Chevron claimed that we had used the plaintiffs' distress to extort money from them.
THE TRUE ADMINISTRATION OF JUSTICE IS THE FIRMEST PILLAR OF GOOD GOVERNMENT
Chevron's lawyers invoked the RICO Act, which was passed in 1970 to combat the Mafia. It meant that all the members of an organization could be indicted.
Steven Donziger and I were the main accused.
Everyone who'd helped us was attacked for being complicit...
...sponsors, NGOs, even mere students who'd done placements with us.
The case arrived on the desk of Judge Lewis Kaplan.

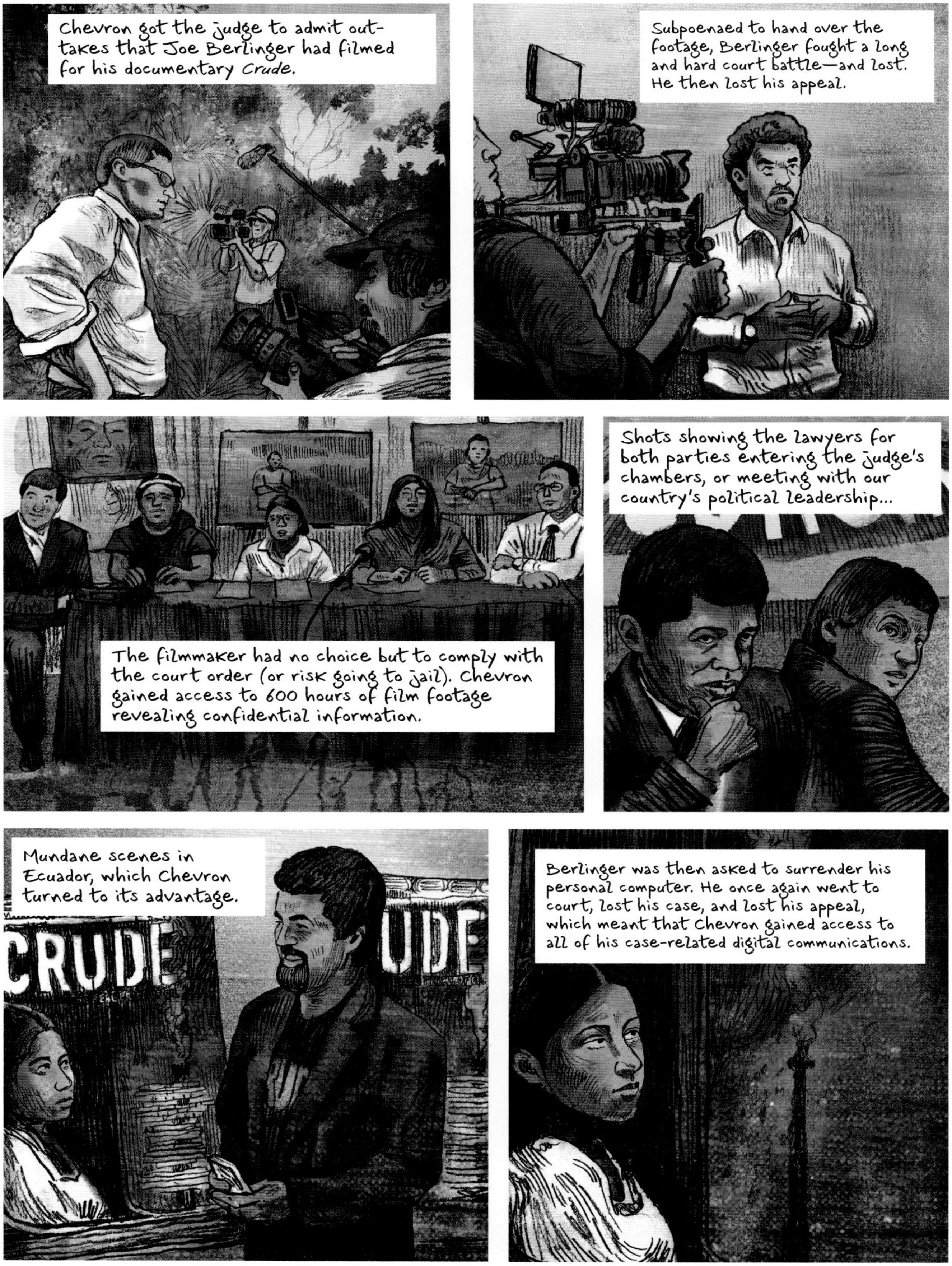
Chevron got the judge to admit out-takes that Joe Berlinger had filmed for his documentary *Crude*.
Subpoenaed to hand over the footage, Berlinger fought a long and hard court battle—and lost. He then lost his appeal.
The filmmaker had no choice but to comply with the court order (or risk going to jail). Chevron gained access to 600 hours of film footage revealing confidential information.
Shots showing the lawyers for both parties entering the judge's chambers, or meeting with our country's political leadership...
Mundane scenes in Ecuador, which Chevron turned to its advantage.
CRUDE
UDE
Berlinger was then asked to surrender his personal computer. He once again went to court, lost his case, and lost his appeal, which meant that Chevron gained access to all of his case-related digital communications.

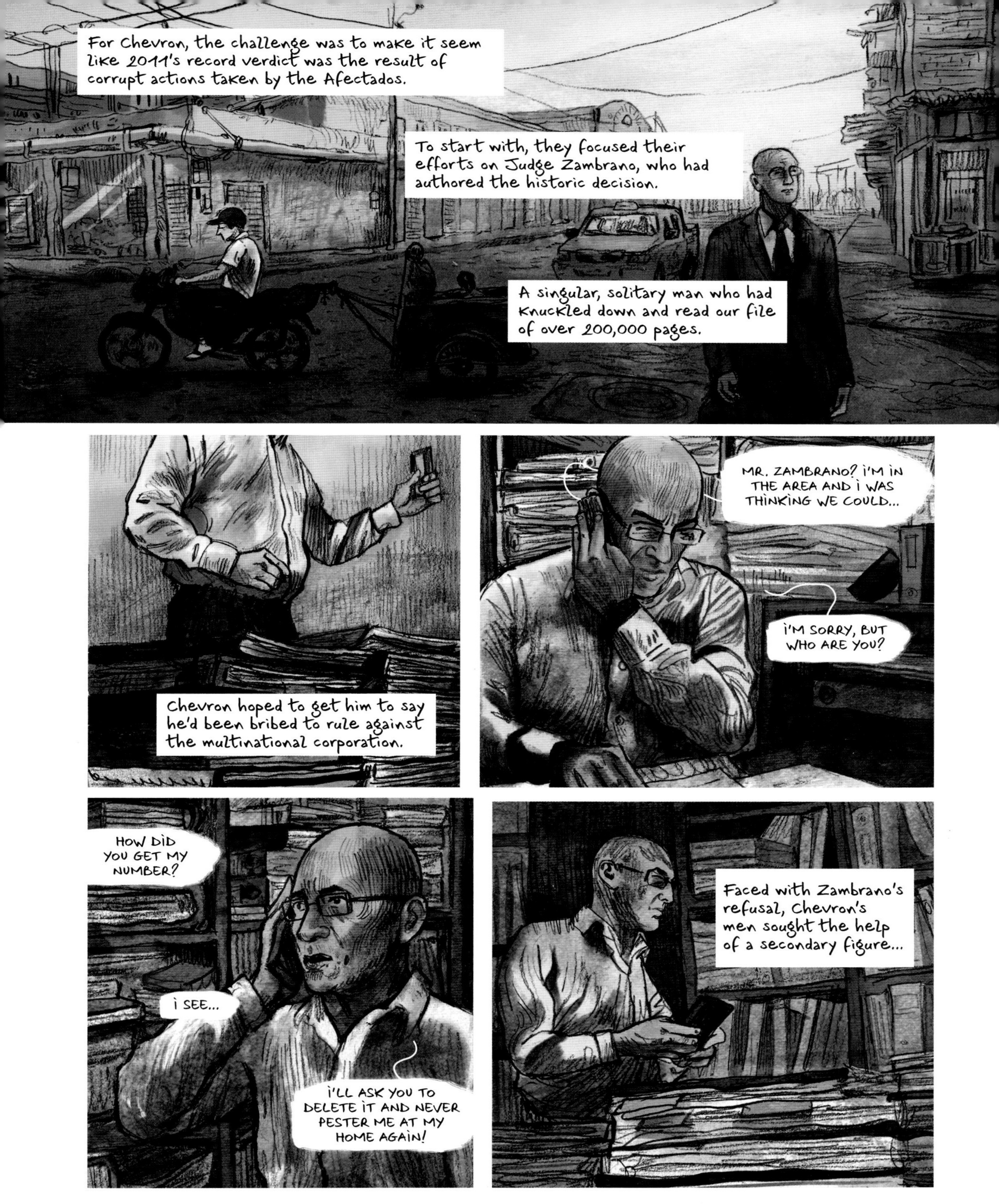
For Chevron, the challenge was to make it seem like 2011's record verdict was the result of corrupt actions taken by the Afectados.
To start with, they focused their efforts on Judge Zambrano, who had authored the historic decision.
A singular, solitary man who had knuckled down and read our file of over 200,000 pages.
Chevron hoped to get him to say he'd been bribed to rule against the multinational corporation.
MR. ZAMBRANO? I'M IN THE AREA AND I WAS THINKING WE COULD...
I'M SORRY, BUT WHO ARE YOU?
HOW DID YOU GET MY NUMBER?
I SEE...
I'LL ASK YOU TO DELETE IT AND NEVER PESTER ME AT MY HOME AGAIN!
Faced with Zambrano's refusal, Chevron's men sought the help of a secondary figure...

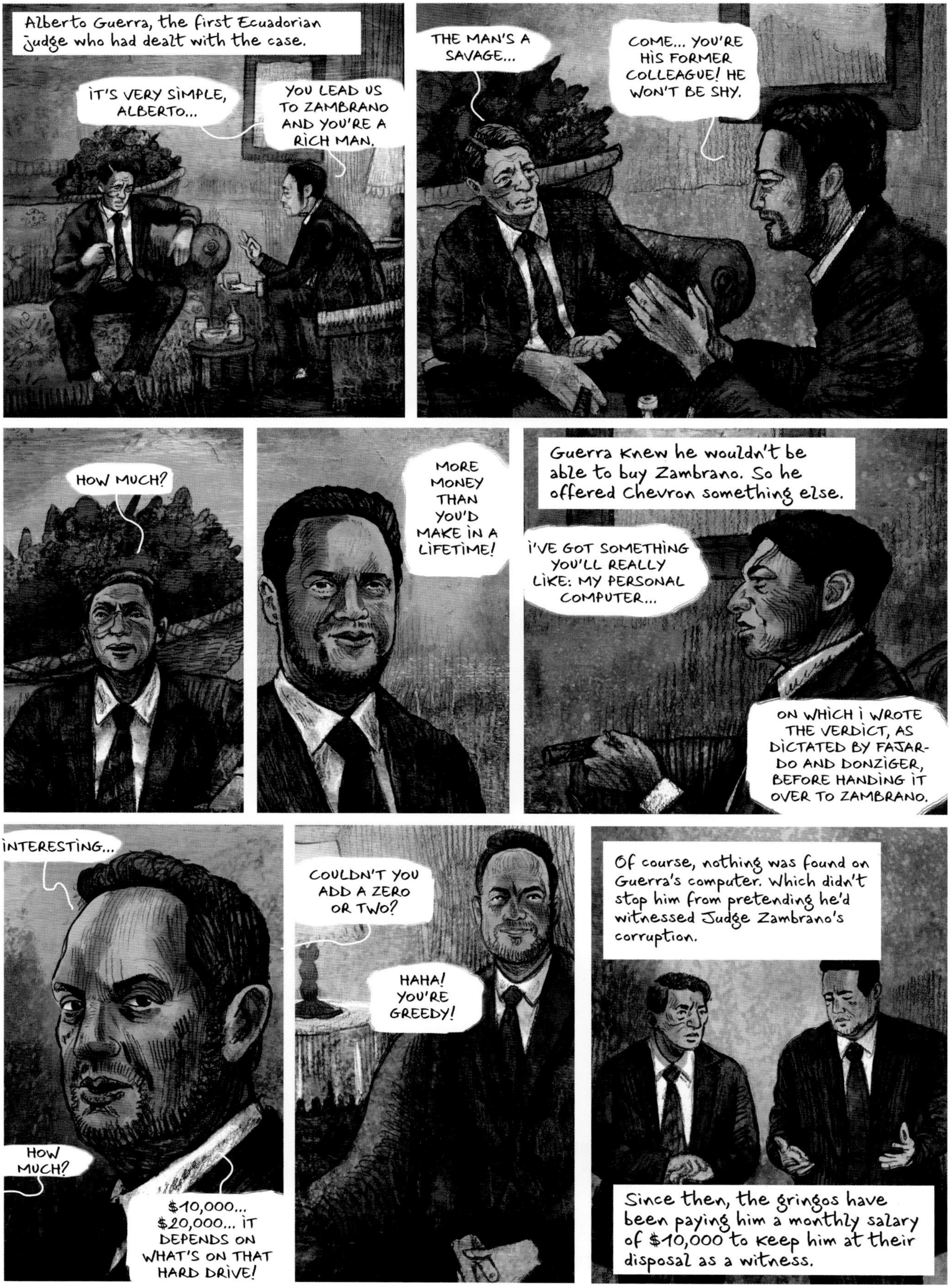

Alberto Guerra, the first Ecuadorian judge who had dealt with the case.
IT'S VERY SIMPLE, ALBERTO...
YOU LEAD US TO ZAMBRANO AND YOU'RE A RICH MAN.
THE MAN'S A SAVAGE...
COME... YOU'RE HIS FORMER COLLEAGUE! HE WON'T BE SHY.
HOW MUCH?
MORE MONEY THAN YOU'D MAKE IN A LIFETIME!
Guerra knew he wouldn't be able to buy Zambrano. So he offered Chevron something else.
I'VE GOT SOMETHING YOU'LL REALLY LIKE: MY PERSONAL COMPUTER...
ON WHICH I WROTE THE VERDICT, AS DICTATED BY FAJARDO AND DONZIGER, BEFORE HANDING IT OVER TO ZAMBRANO.
INTERESTING...
HOW MUCH?
$10,000... $20,000... IT DEPENDS ON WHAT'S ON THAT HARD DRIVE!
COULDN'T YOU ADD A ZERO OR TWO?
HAHA! YOU'RE GREEDY!
Of course, nothing was found on Guerra's computer. Which didn't stop him from pretending he'd witnessed Judge Zambrano's corruption.
Since then, the gringos have been paying him a monthly salary of $10,000 to keep him at their disposal as a witness.

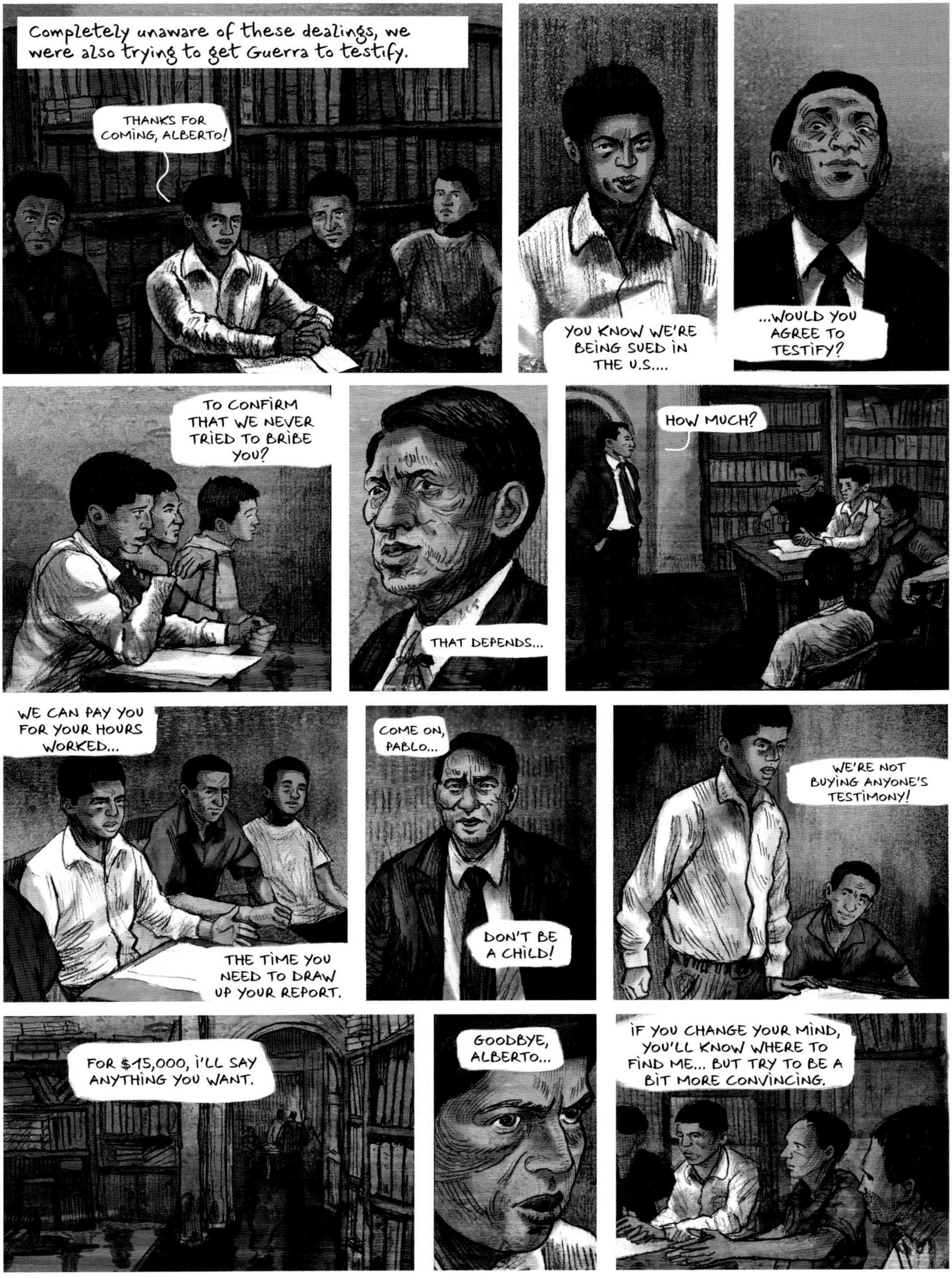
Completely unaware of these dealings, we were also trying to get Guerra to testify.
THANKS FOR COMING, ALBERTO!
YOU KNOW WE'RE BEING SUED IN THE U.S....
...WOULD YOU AGREE TO TESTIFY?
TO CONFIRM THAT WE NEVER TRIED TO BRIBE YOU?
THAT DEPENDS...
HOW MUCH?
WE CAN PAY YOU FOR YOUR HOURS WORKED...
THE TIME YOU NEED TO DRAW UP YOUR REPORT.
COME ON, PABLO...
DON'T BE A CHILD!
WE'RE NOT BUYING ANYONE'S TESTIMONY!
FOR $15,000, I'LL SAY ANYTHING YOU WANT.
GOODBYE, ALBERTO...
IF YOU CHANGE YOUR MIND, YOU'LL KNOW WHERE TO FIND ME... BUT TRY TO BE A BIT MORE CONVINCING.

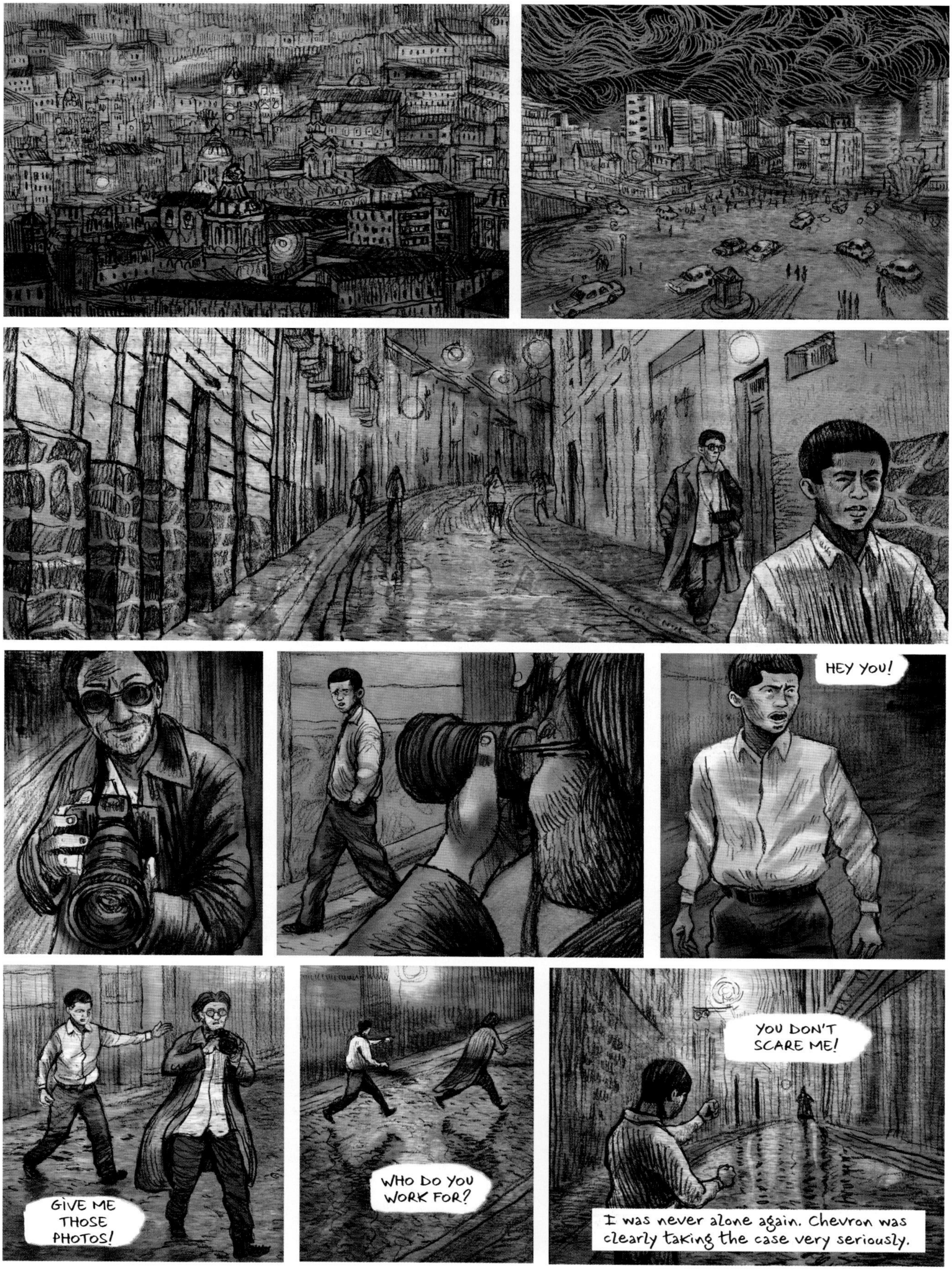
HEY YOU!
GIVE ME THOSE PHOTOS!
WHO DO YOU WORK FOR?
YOU DON'T SCARE ME!
I was never alone again. Chevron was clearly taking the case very seriously.

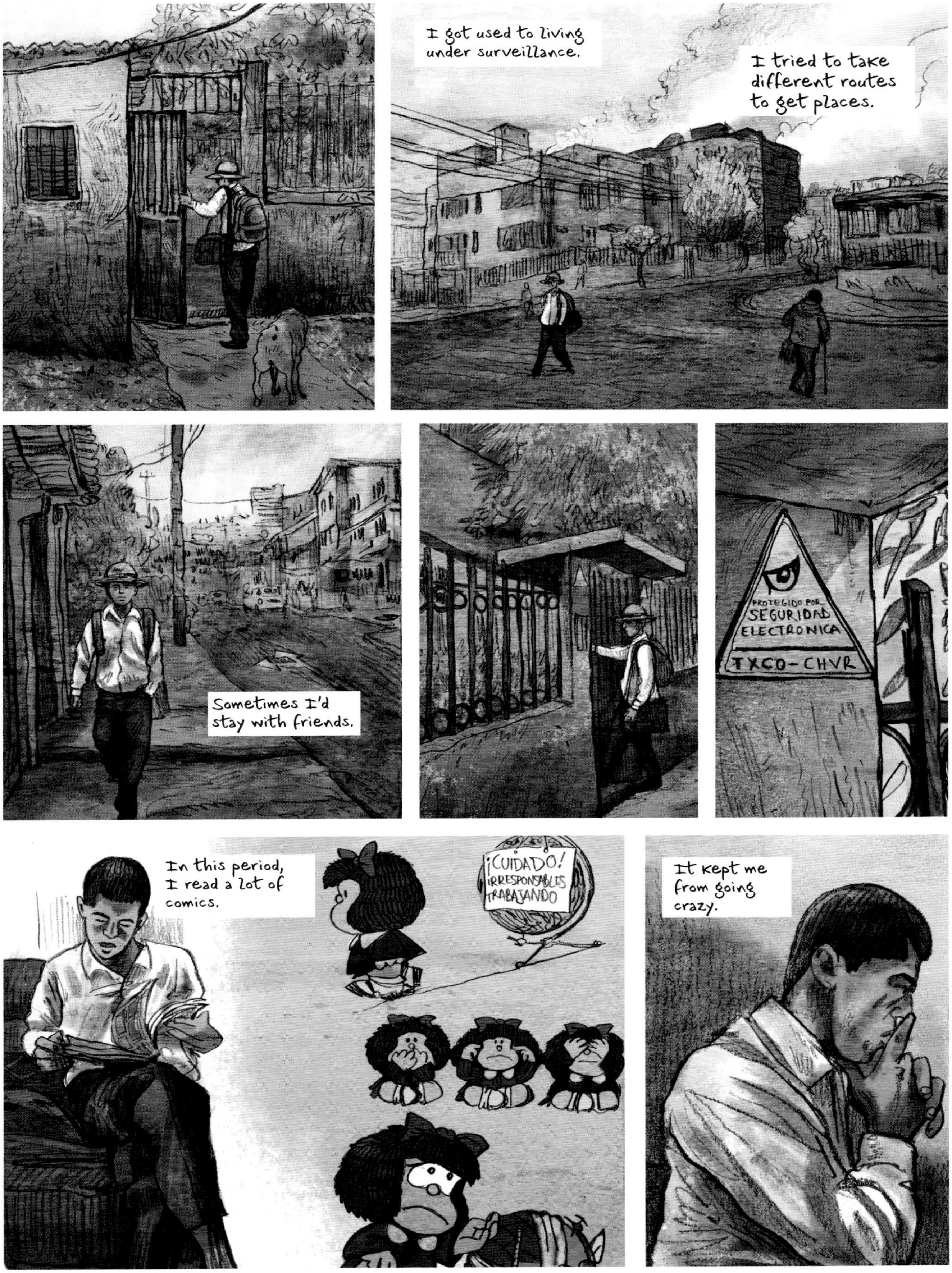
I got used to living under surveillance.
I tried to take different routes to get places.
Sometimes I'd stay with friends.
PROTEGIDO POR SEGURIDAD ELECTRONICA
TXCO-CHVR
In this period, I read a lot of comics.
¡CUIDADO! IRRESPONSABLES TRABAJANDO
It kept me from going crazy.

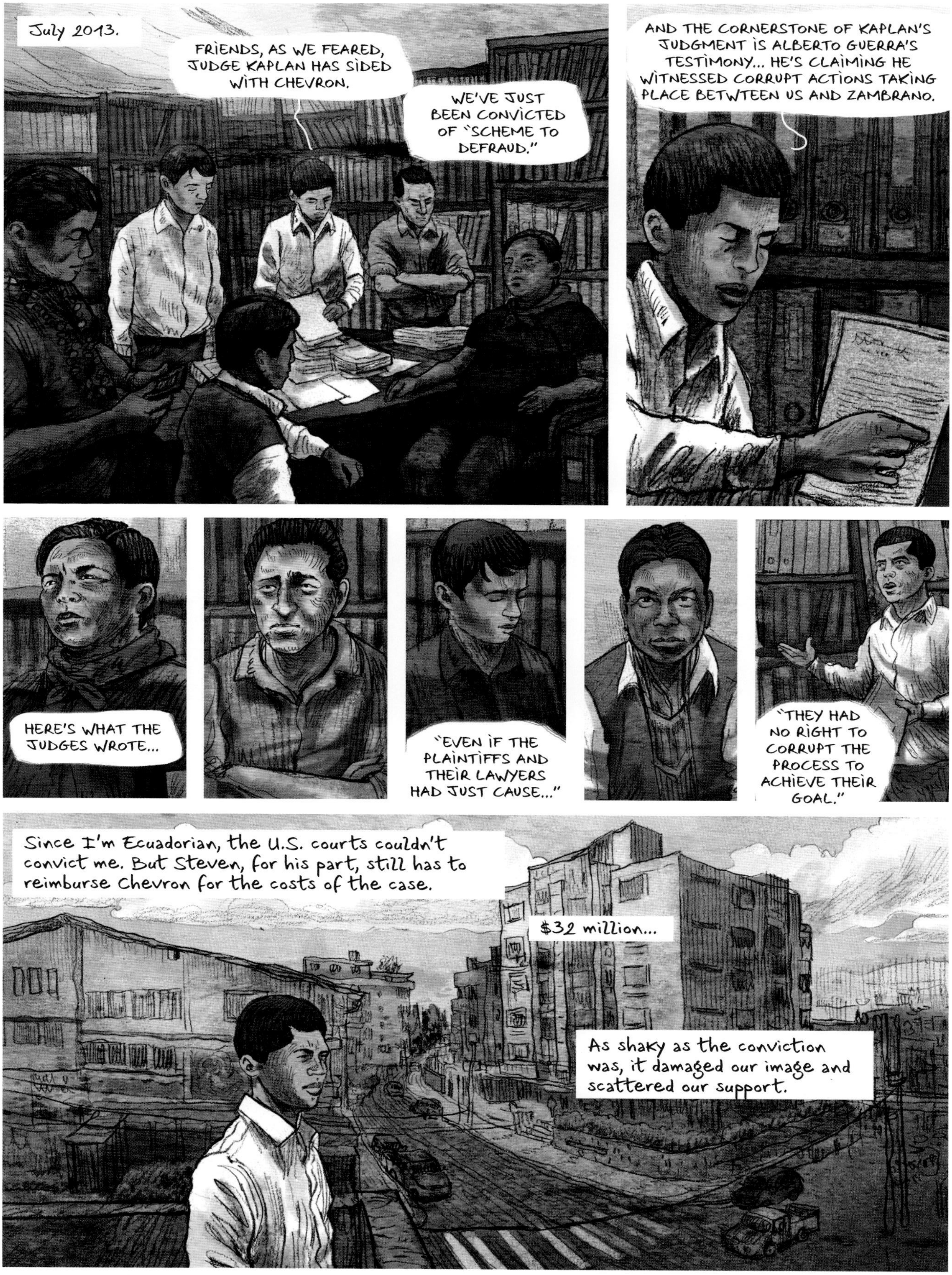
July 2013.
FRIENDS, AS WE FEARED, JUDGE KAPLAN HAS SIDED WITH CHEVRON.
WE'VE JUST BEEN CONVICTED OF "SCHEME TO DEFRAUD."
AND THE CORNERSTONE OF KAPLAN'S JUDGMENT IS ALBERTO GUERRA'S TESTIMONY... HE'S CLAIMING HE WITNESSED CORRUPT ACTIONS TAKING PLACE BETWTEEN US AND ZAMBRANO.
HERE'S WHAT THE JUDGES WROTE...
"EVEN IF THE PLAINTIFFS AND THEIR LAWYERS HAD JUST CAUSE..."
"THEY HAD NO RIGHT TO CORRUPT THE PROCESS TO ACHIEVE THEIR GOAL."
Since I'm Ecuadorian, the U.S. courts couldn't convict me. But Steven, for his part, still has to reimburse Chevron for the costs of the case.
$32 million...
As shaky as the conviction was, it damaged our image and scattered our support.

To raise money, we started organizing auctions and bazaars, craft fairs...

I could no longer go without pay.

In April 2015, Alberto Guerra, at a hearing in Washington for a case against Chevron and the government of Ecuador, admitted to having given false testimony on Chevron's behalf.

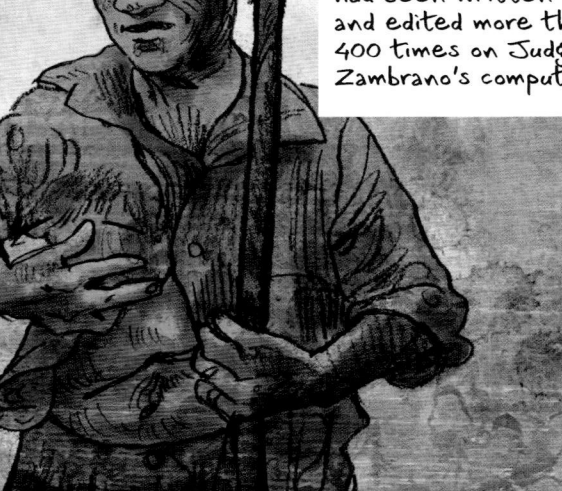

Experts also showed that, contrary to what Guerra claimed, the 2011 verdict had been written and edited more than 400 times on Judge Zambrano's computer.

This was not enough to change the course of justice in the U.S.

In July 2015, the Court of Appeals in New York upheld the lower court's ruling.

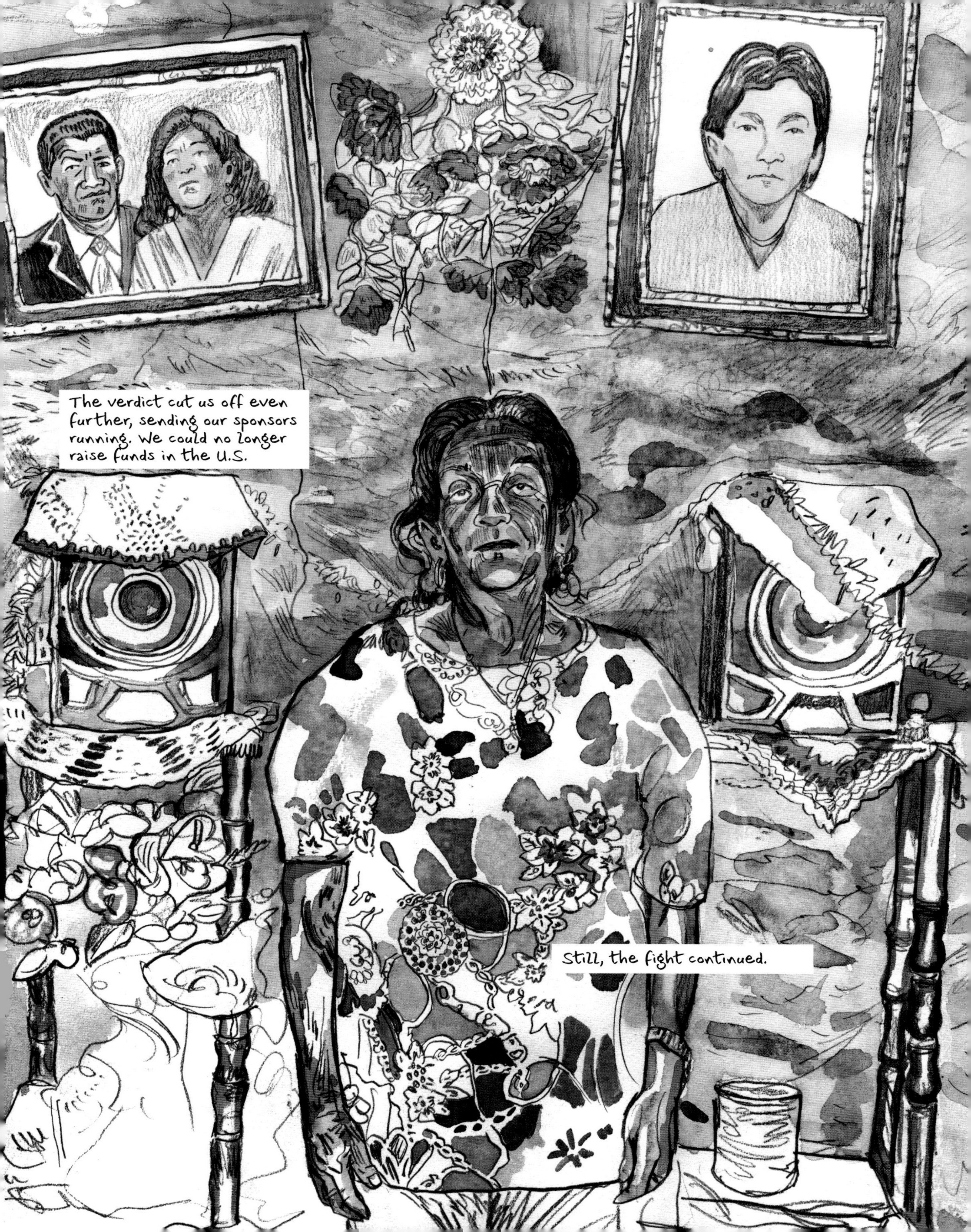
The verdict cut us off even further, sending our sponsors running. We could no longer raise funds in the U.S.
Still, the fight continued.

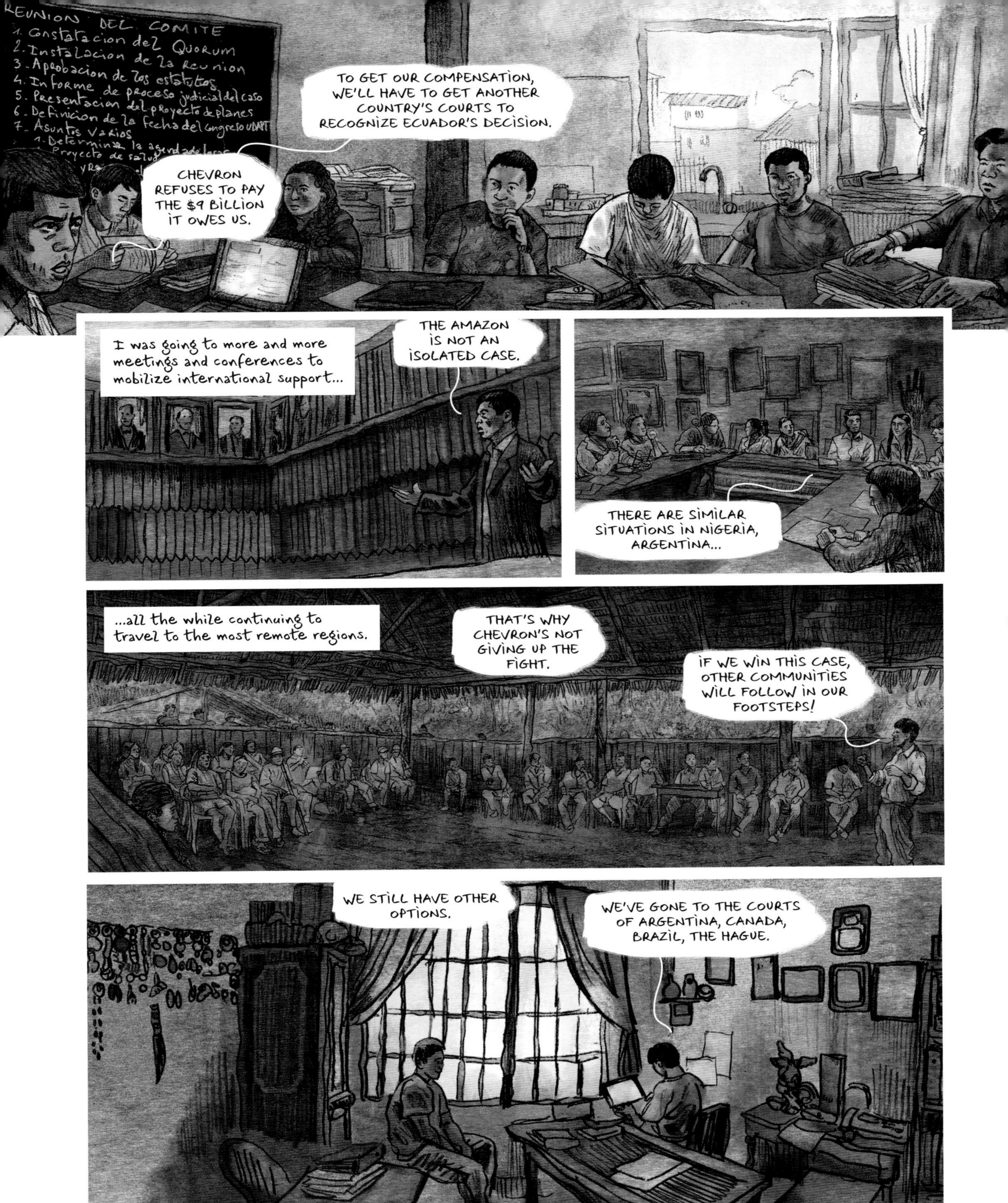
REUNION DEL COMITE
1. Constatacion del Quorum
2. Instalacion de la Reunion
3. Aprobacion de los estatutos
4. Informe de proceso judicial del caso
5. Presentacion del proyecto de planes
6. Definicion de la fecha del congreso UDAPT
7. Asuntos Varios
TO GET OUR COMPENSATION, WE'LL HAVE TO GET ANOTHER COUNTRY'S COURTS TO RECOGNIZE ECUADOR'S DECISION.
CHEVRON REFUSES TO PAY THE $9 BILLION IT OWES US.
I was going to more and more meetings and conferences to mobilize international support...
THE AMAZON IS NOT AN ISOLATED CASE.
THERE ARE SIMILAR SITUATIONS IN NIGERIA, ARGENTINA...
...all the while continuing to travel to the most remote regions.
THAT'S WHY CHEVRON'S NOT GIVING UP THE FIGHT.
IF WE WIN THIS CASE, OTHER COMMUNITIES WILL FOLLOW IN OUR FOOTSTEPS!
WE STILL HAVE OTHER OPTIONS.
WE'VE GONE TO THE COURTS OF ARGENTINA, CANADA, BRAZIL, THE HAGUE.

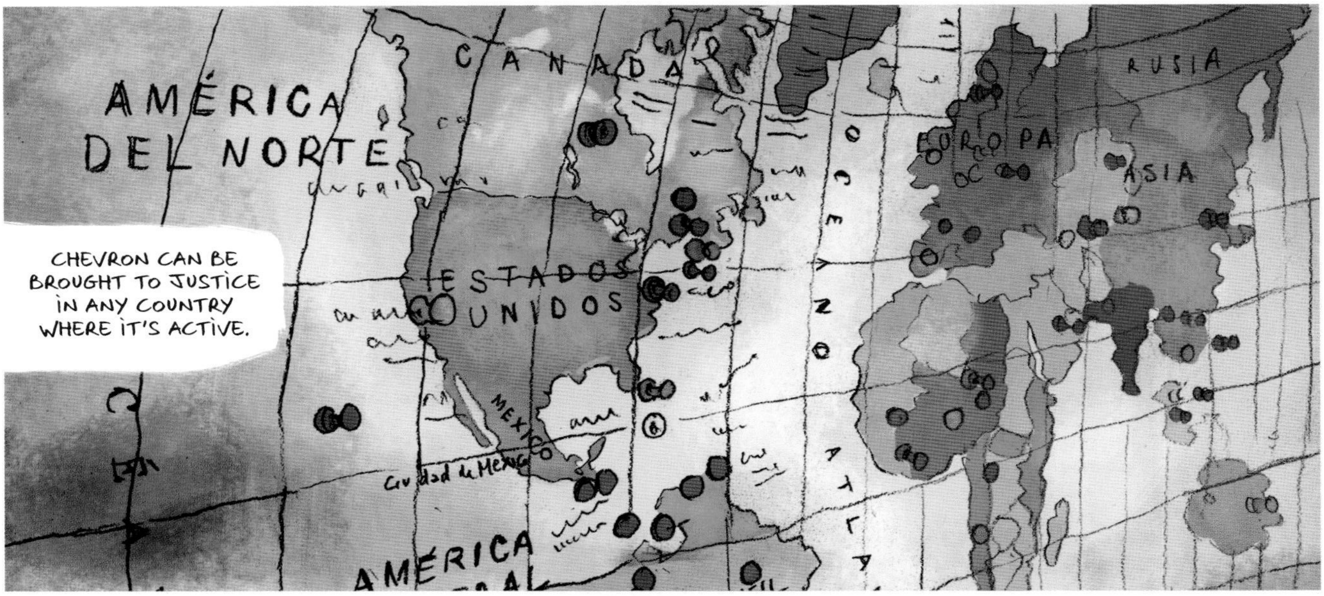
AMÉRICA DEL NORTE
CANADA
RUSIA
EUROPA
ASIA
ESTADOS UNIDOS
MEXICO
OCEANO ATLA
AMÉRICA
CHEVRON CAN BE BROUGHT TO JUSTICE IN ANY COUNTRY WHERE IT'S ACTIVE.

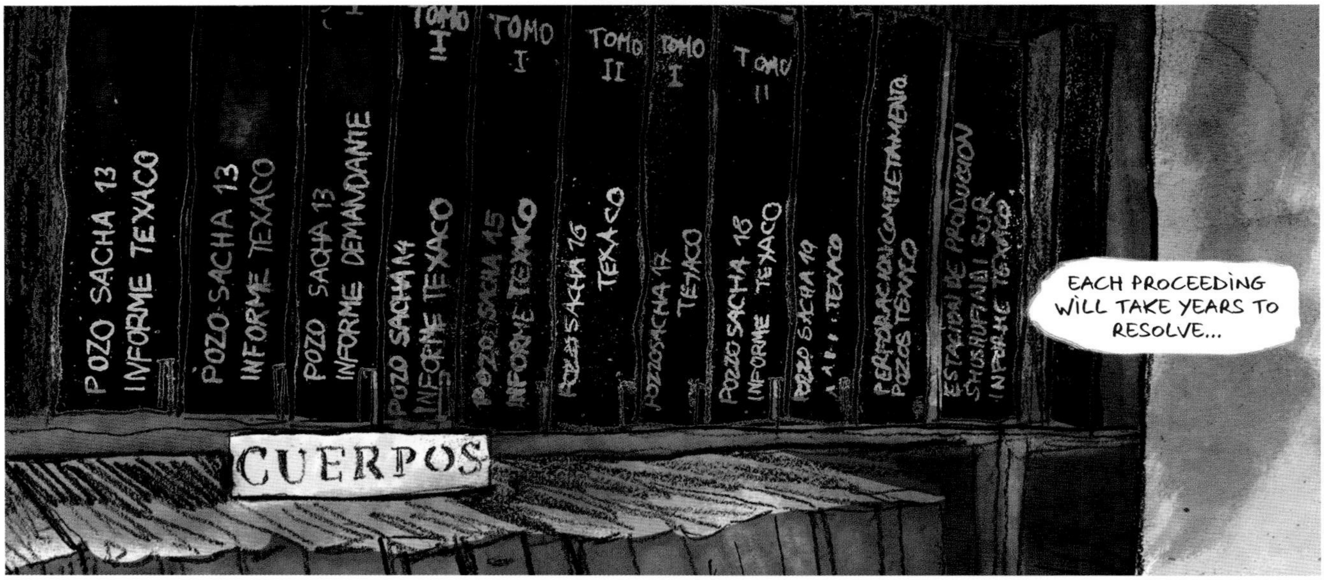
POZO SACHA 13 INFORME TEXACO
POZO SACHA 13 INFORME TEXACO
POZO SACHA 13 INFORME DEMANDANTE
CUERPOS
EACH PROCEEDING WILL TAKE YEARS TO RESOLVE...

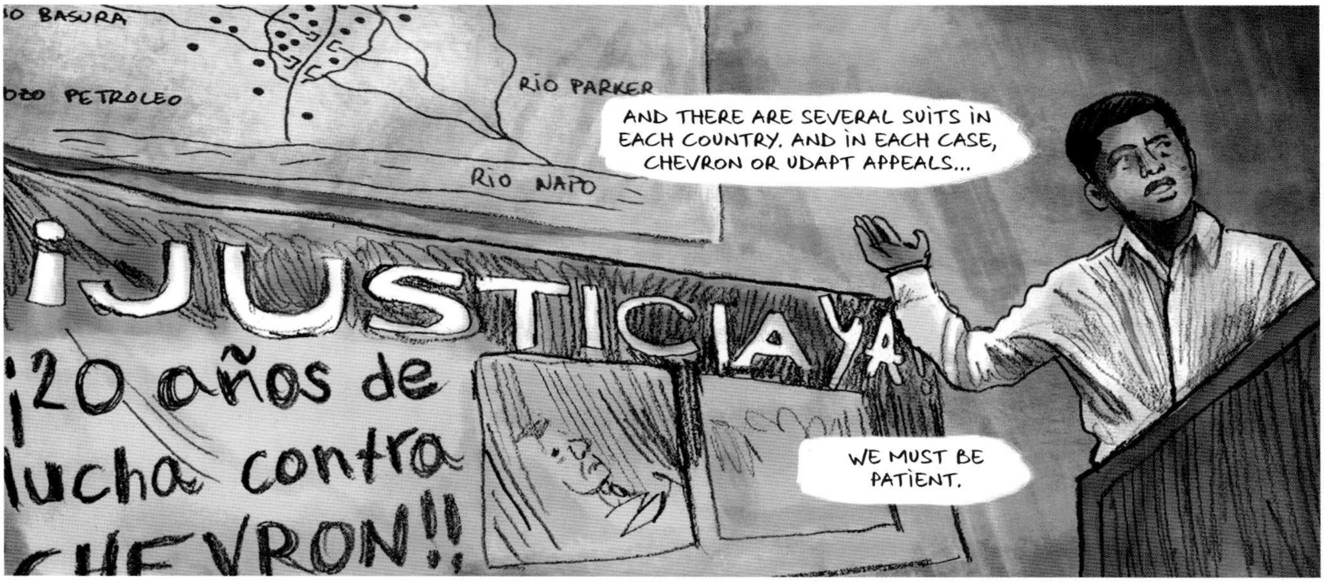
RIO PARKER
RIO NAPO
¡JUSTICIA YA
¡20 años de lucha contra CHEVRON!!
AND THERE ARE SEVERAL SUITS IN EACH COUNTRY. AND IN EACH CASE, CHEVRON OR UDAPT APPEALS...
WE MUST BE PATIENT.

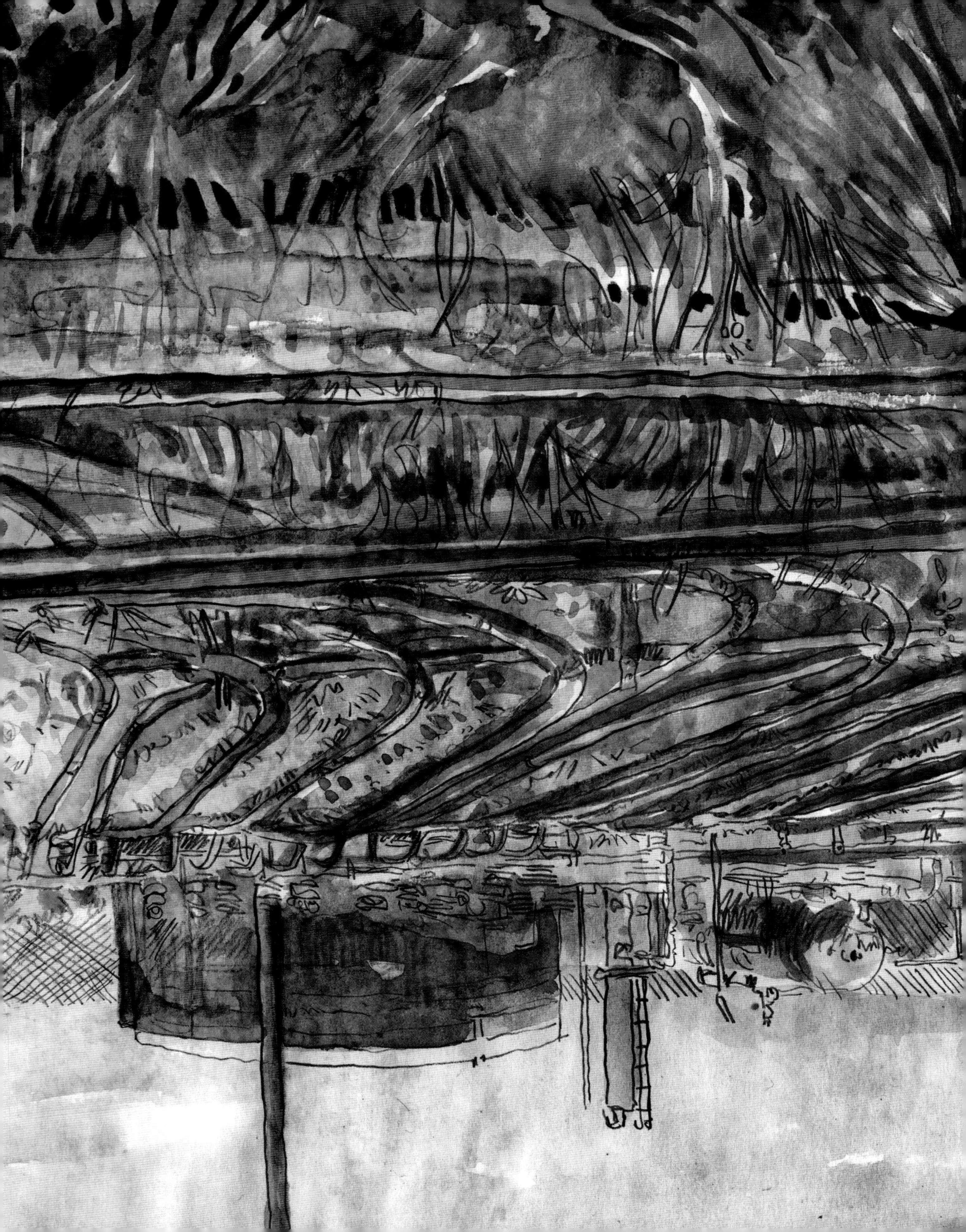

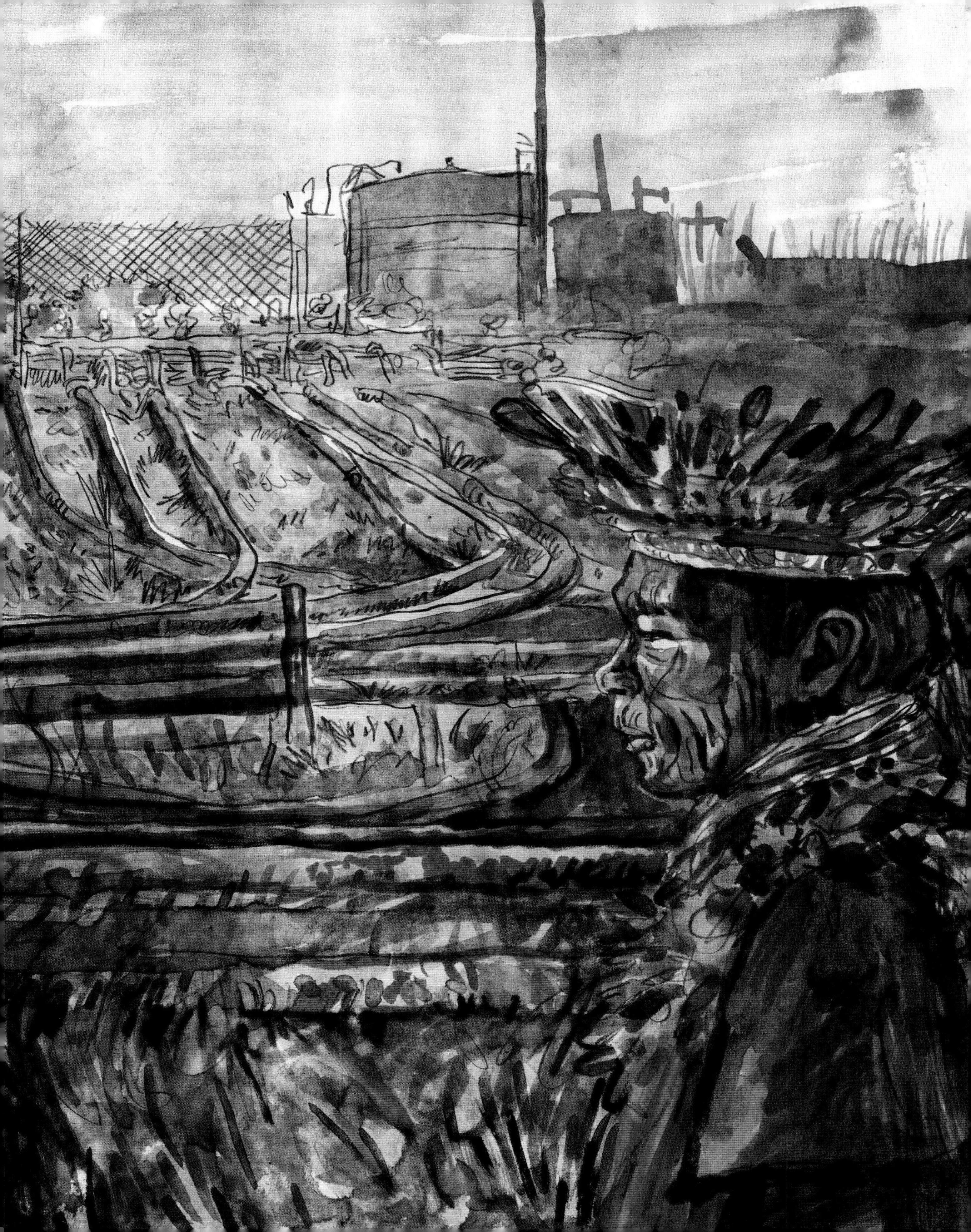

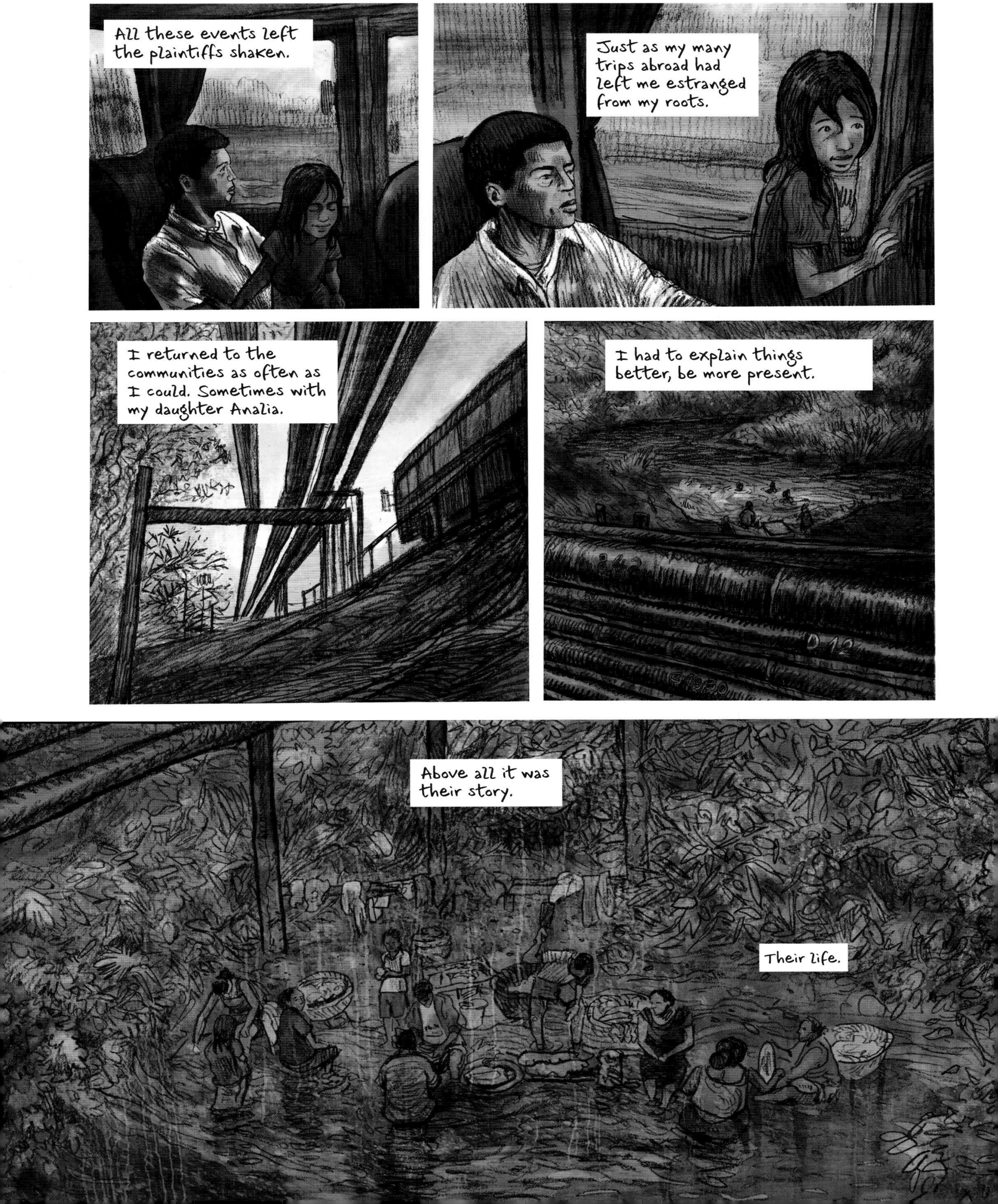
All these events left the plaintiffs shaken.
Just as my many trips abroad had left me estranged from my roots.
I returned to the communities as often as I could. Sometimes with my daughter Analia.
I had to explain things better, be more present.
D 12
Above all it was their story.
Their life.

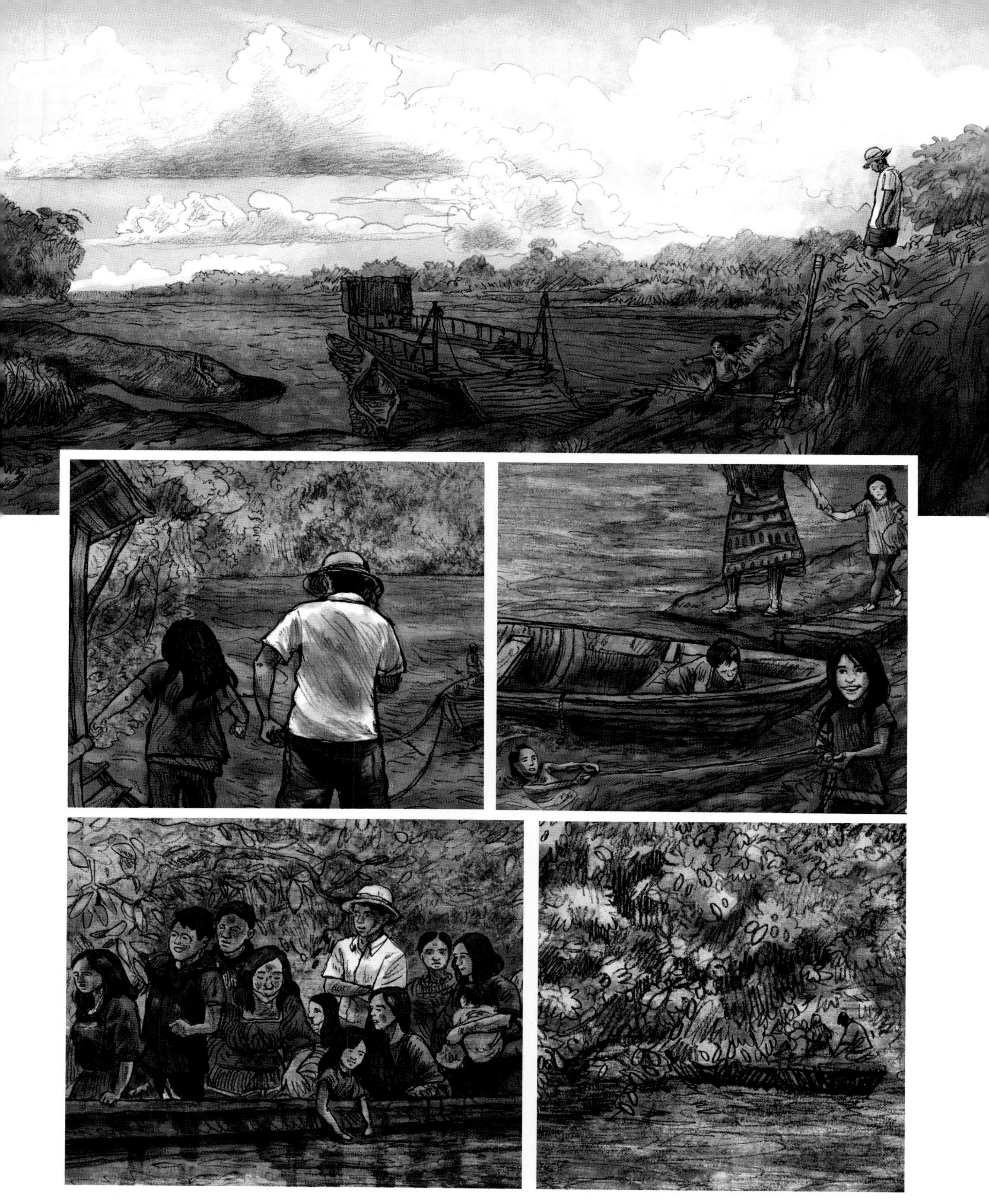

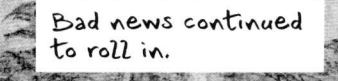
Bad news continued to roll in.

The appeals we'd made to other countries were floundering.

One after another, the courts declined jurisdiction.

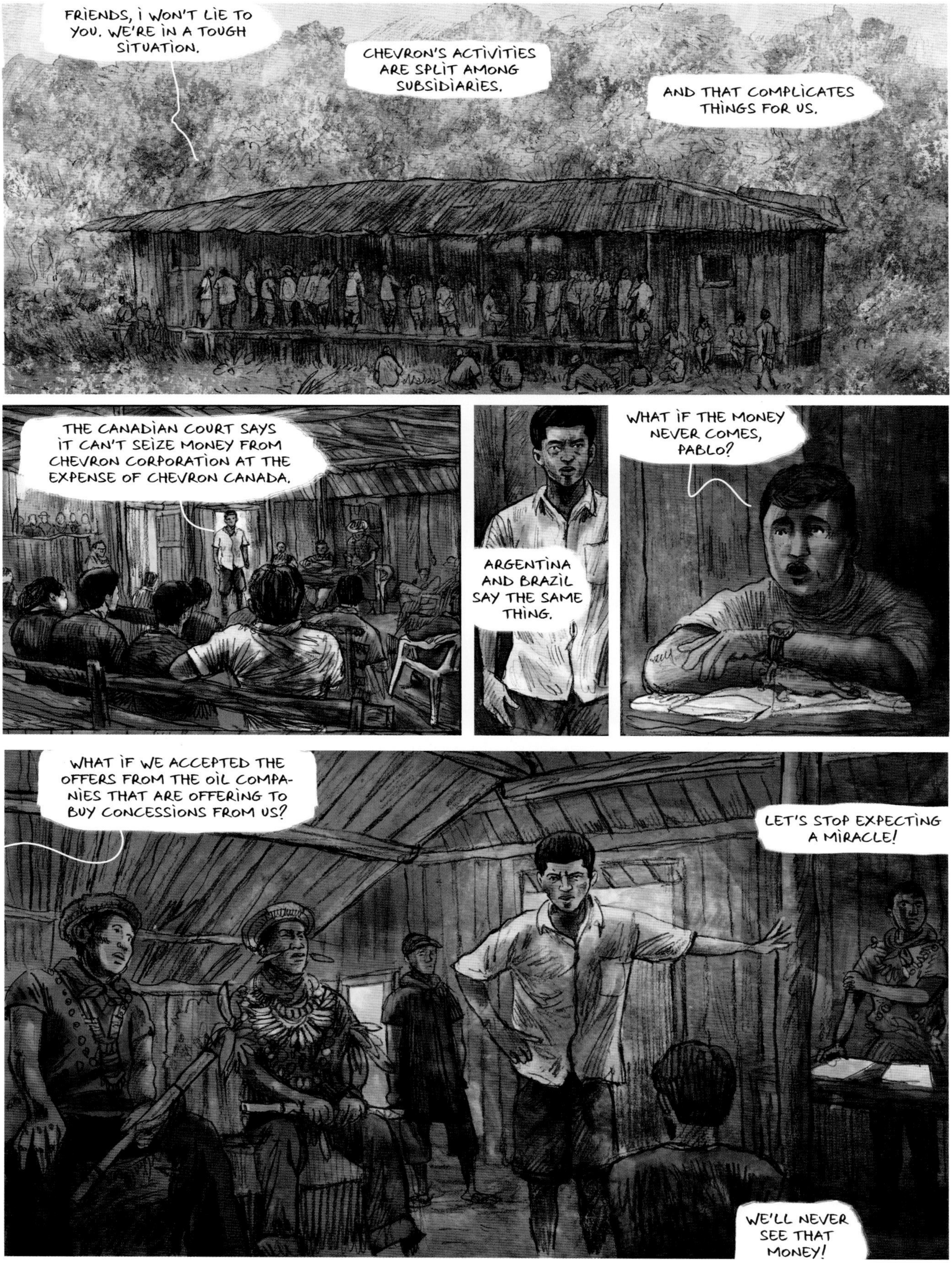
FRIENDS, I WON'T LIE TO YOU. WE'RE IN A TOUGH SITUATION.
CHEVRON'S ACTIVITIES ARE SPLIT AMONG SUBSIDIARIES.
AND THAT COMPLICATES THINGS FOR US.
THE CANADIAN COURT SAYS IT CAN'T SEIZE MONEY FROM CHEVRON CORPORATION AT THE EXPENSE OF CHEVRON CANADA.
ARGENTINA AND BRAZIL SAY THE SAME THING.
WHAT IF THE MONEY NEVER COMES, PABLO?
WHAT IF WE ACCEPTED THE OFFERS FROM THE OIL COMPANIES THAT ARE OFFERING TO BUY CONCESSIONS FROM US?
LET'S STOP EXPECTING A MIRACLE!
WE'LL NEVER SEE THAT MONEY!

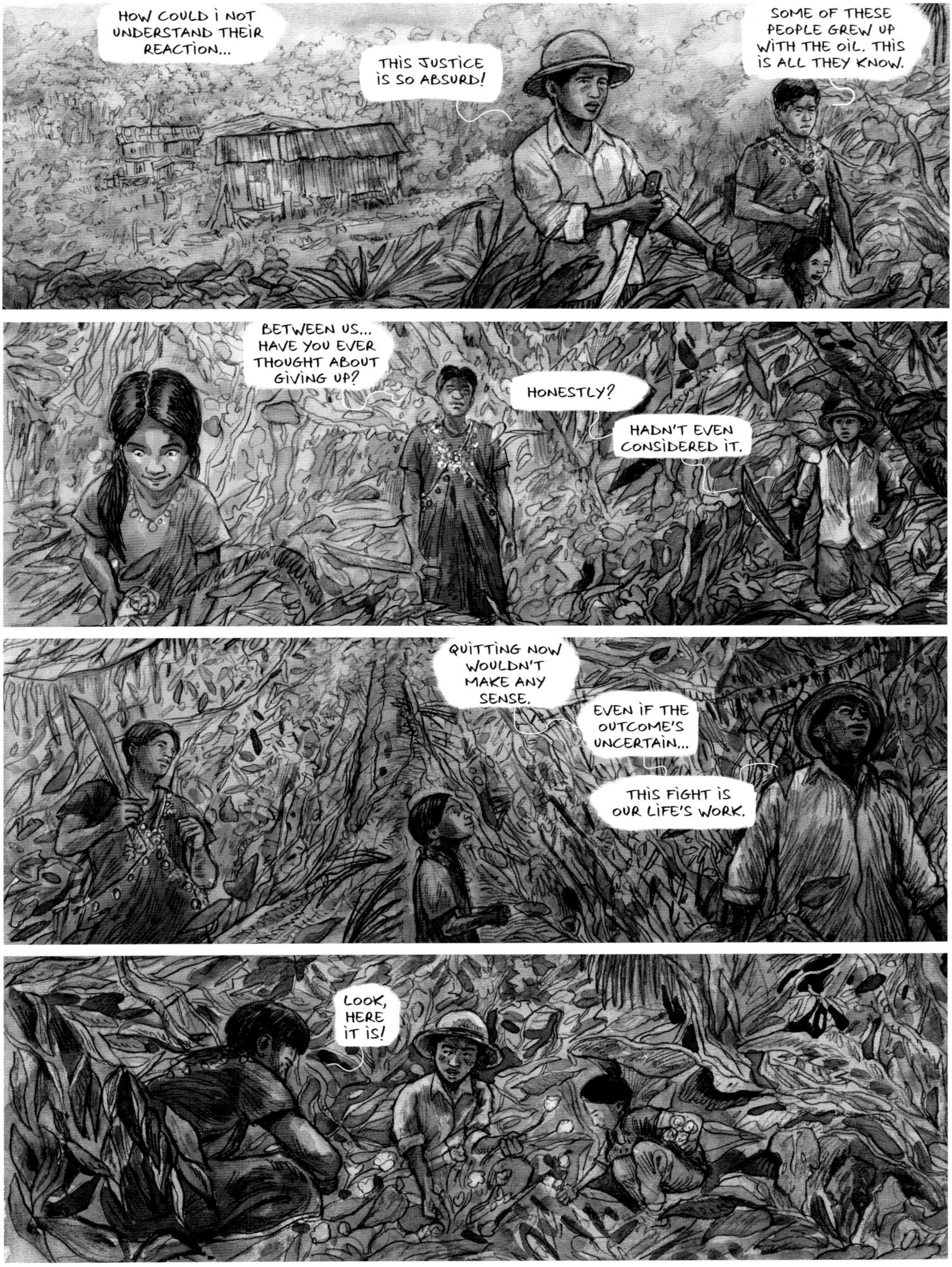
HOW COULD I NOT UNDERSTAND THEIR REACTION...
THIS JUSTICE IS SO ABSURD!
SOME OF THESE PEOPLE GREW UP WITH THE OIL. THIS IS ALL THEY KNOW.
BETWEEN US... HAVE YOU EVER THOUGHT ABOUT GIVING UP?
HONESTLY?
HADN'T EVEN CONSIDERED IT.
QUITTING NOW WOULDN'T MAKE ANY SENSE.
EVEN IF THE OUTCOME'S UNCERTAIN...
THIS FIGHT IS OUR LIFE'S WORK.
LOOK, HERE IT IS!

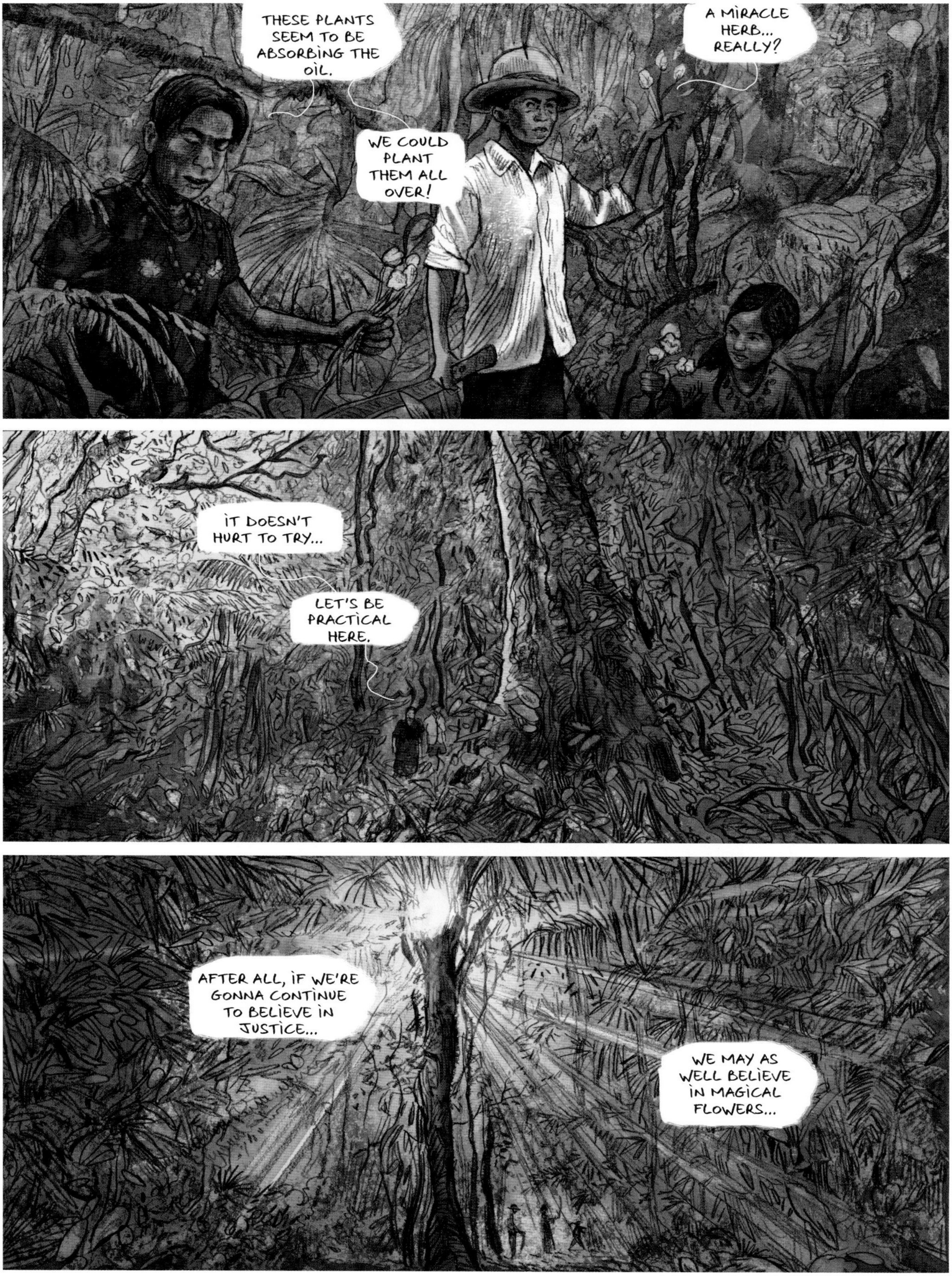
THESE PLANTS SEEM TO BE ABSORBING THE OIL.
WE COULD PLANT THEM ALL OVER!
A MIRACLE HERB... REALLY?
IT DOESN'T HURT TO TRY...
LET'S BE PRACTICAL HERE.
AFTER ALL, IF WE'RE GONNA CONTINUE TO BELIEVE IN JUSTICE...
WE MAY AS WELL BELIEVE IN MAGICAL FLOWERS...

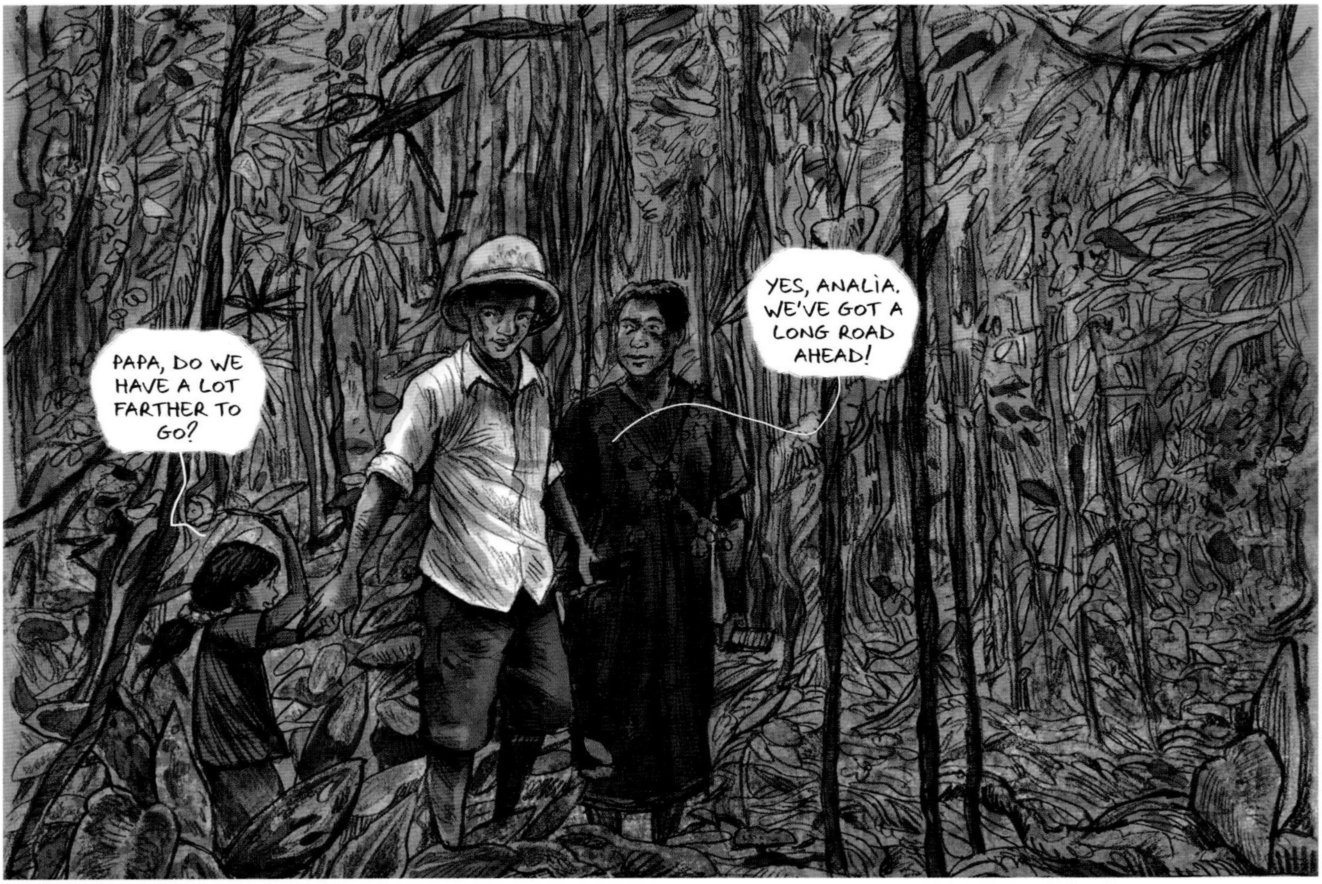
PAPA, DO WE HAVE A LOT FARTHER TO GO?
YES, ANALÌA. WE'VE GOT A LONG ROAD AHEAD!

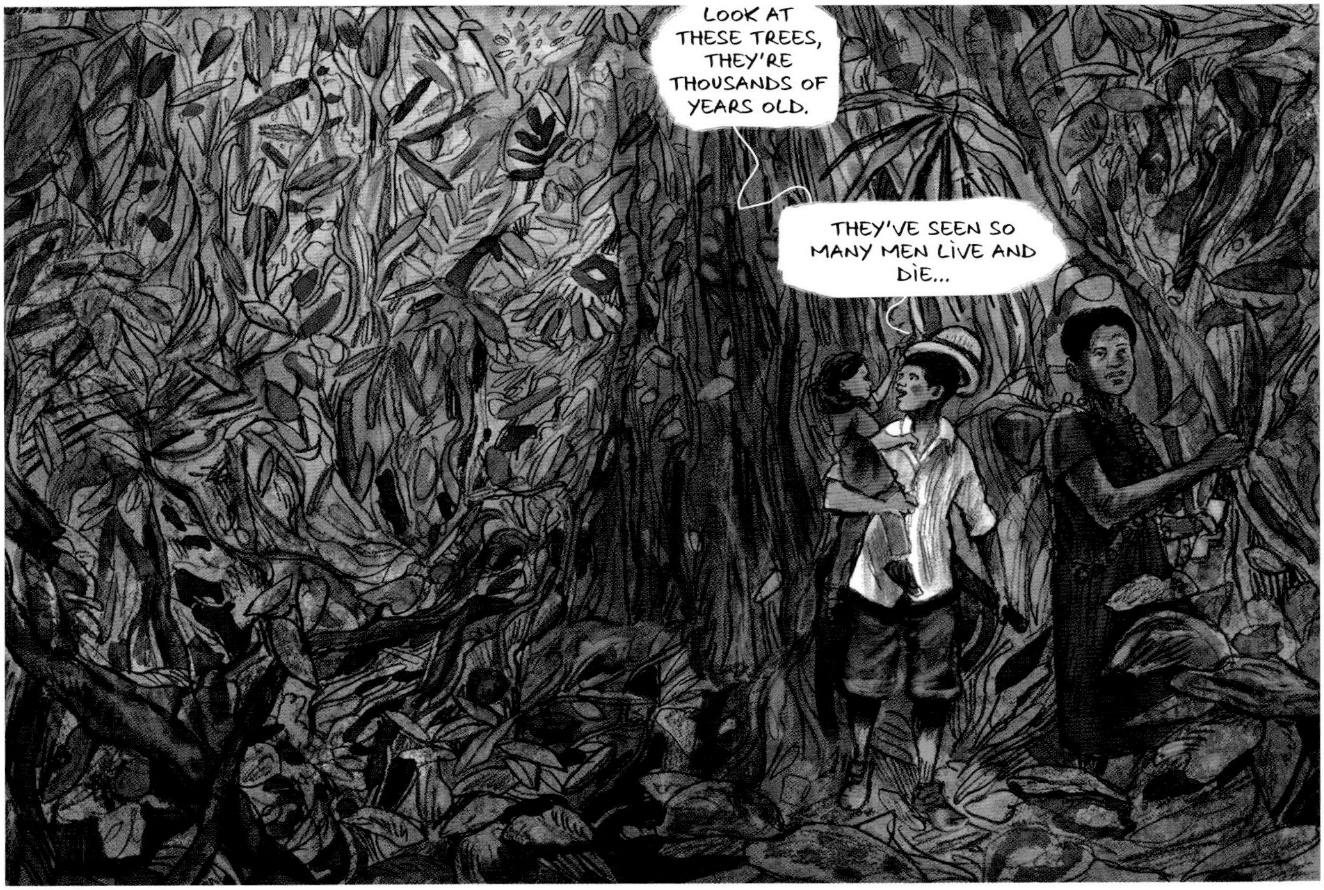
LOOK AT THESE TREES, THEY'RE THOUSANDS OF YEARS OLD.
THEY'VE SEEN SO MANY MEN LÌVE AND DÌE...

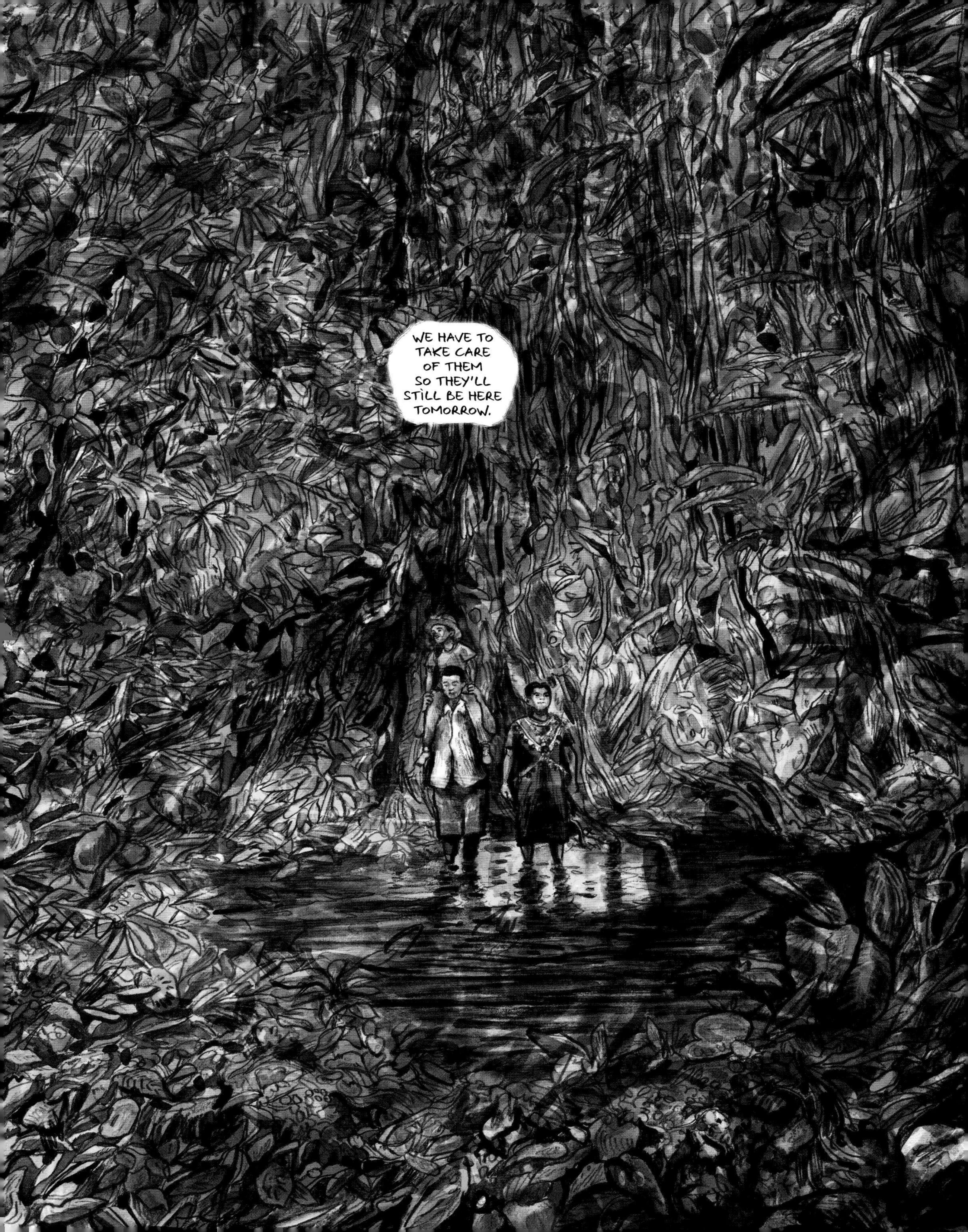
WE HAVE TO TAKE CARE OF THEM SO THEY'LL STILL BE HERE TOMORROW.

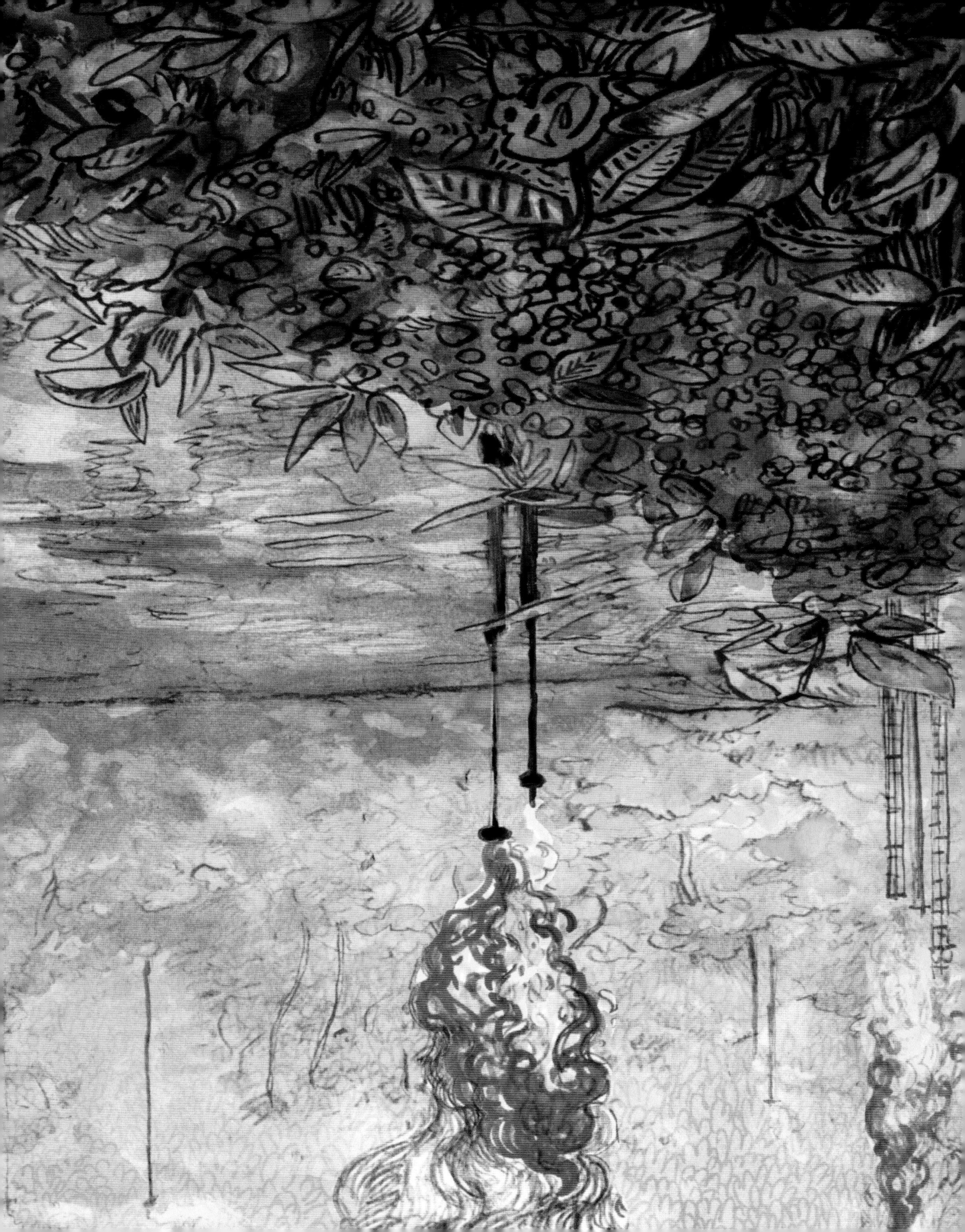

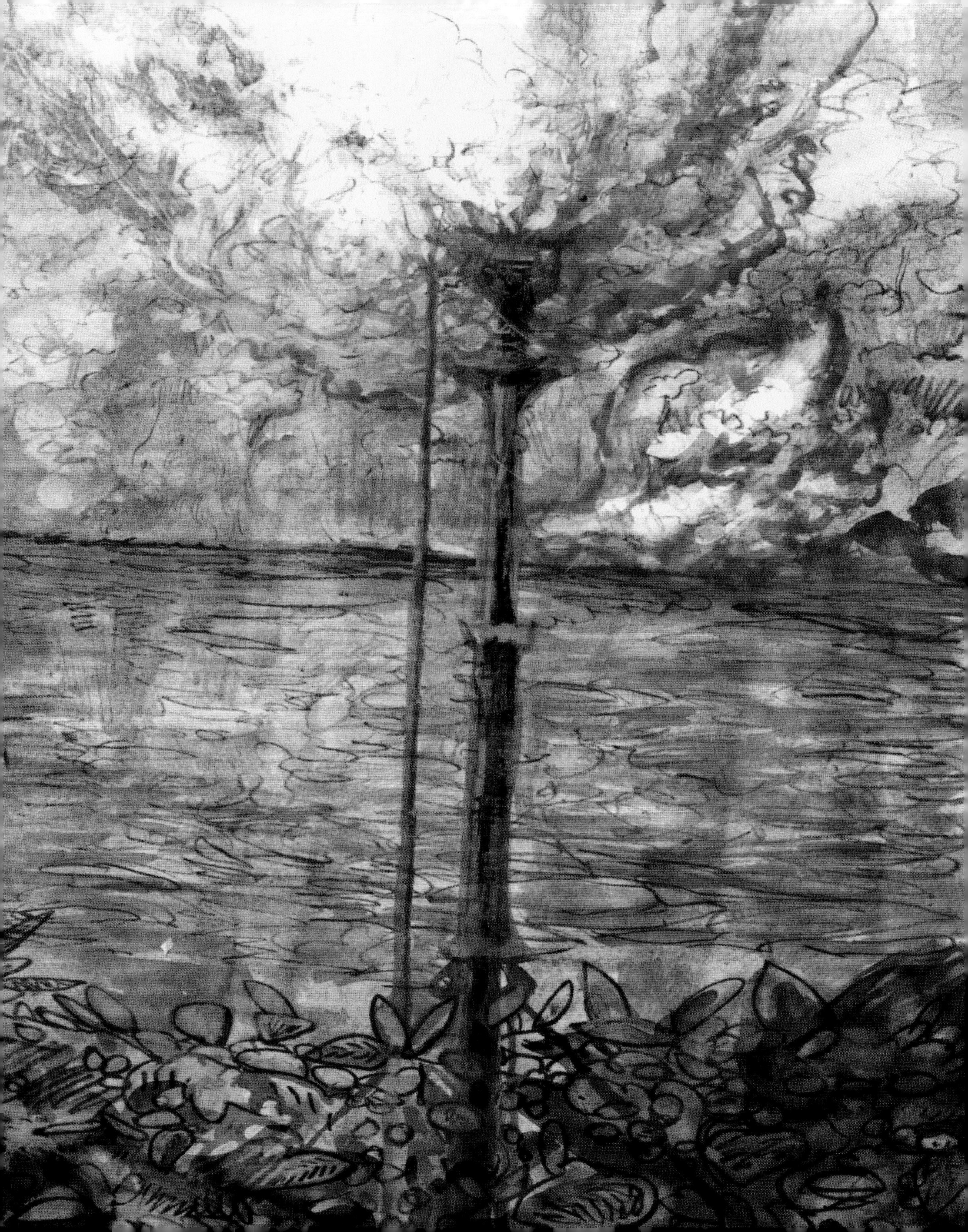

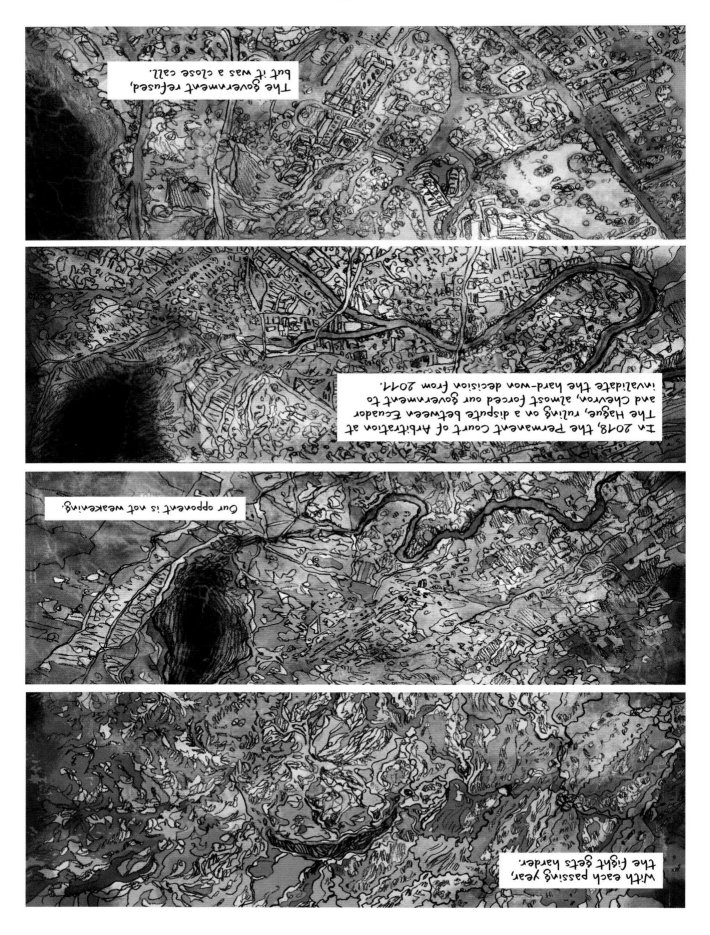

The government refused, but it was a close call.

In 2018, the Permanent Court of Arbitration at The Hague, ruling on a dispute between Ecuador and Chevron, almost forced our government to invalidate the hard-won decision from 2011.

Our opponent is not weakening.

With each passing year, the fight gets harder.

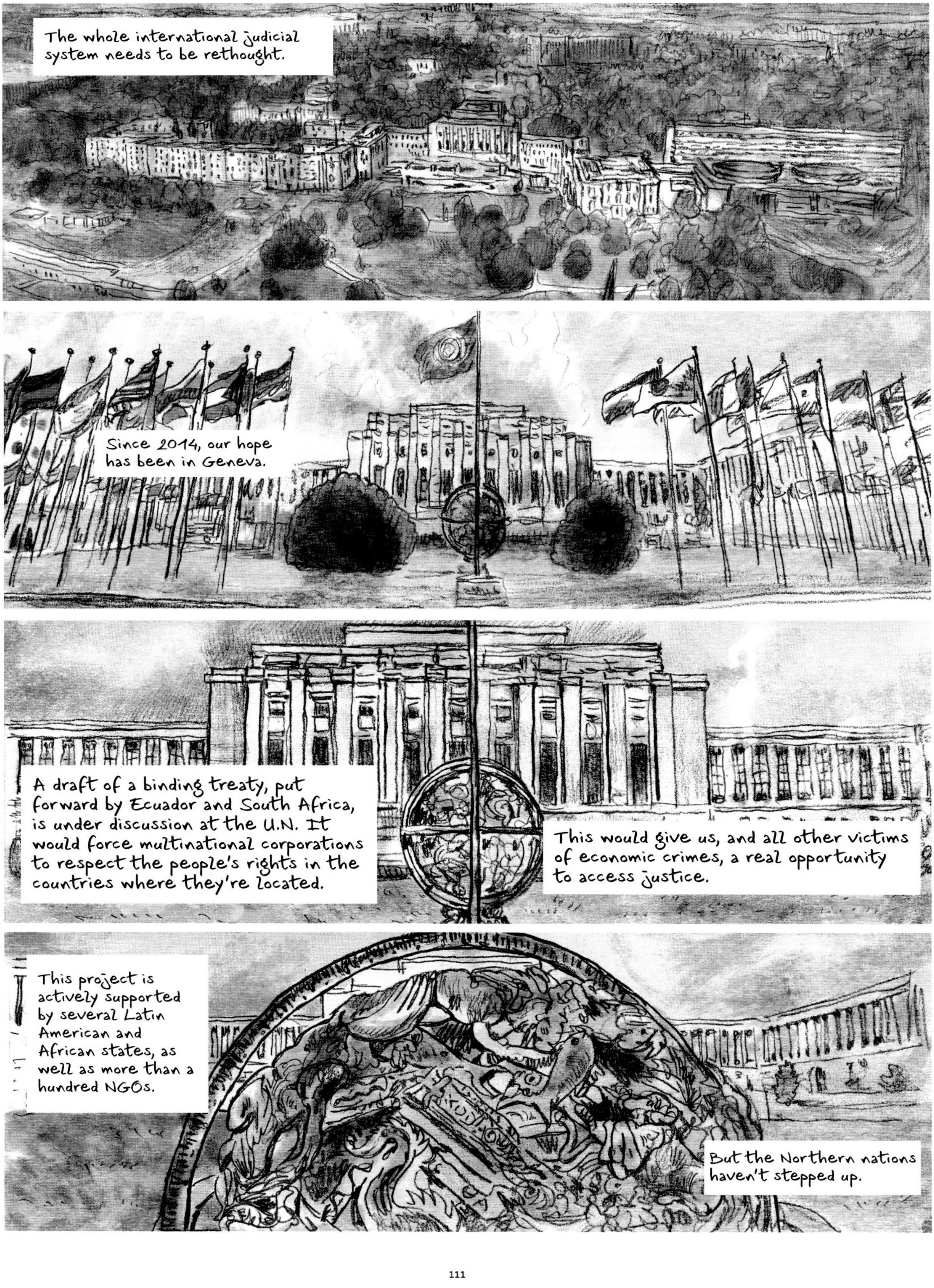
The whole international judicial system needs to be rethought.
Since 2014, our hope has been in Geneva.
A draft of a binding treaty, put forward by Ecuador and South Africa, is under discussion at the U.N. It would force multinational corporations to respect the people's rights in the countries where they're located.
This would give us, and all other victims of economic crimes, a real opportunity to access justice.
This project is actively supported by several Latin American and African states, as well as more than a hundred NGOs.
But the Northern nations haven't stepped up.

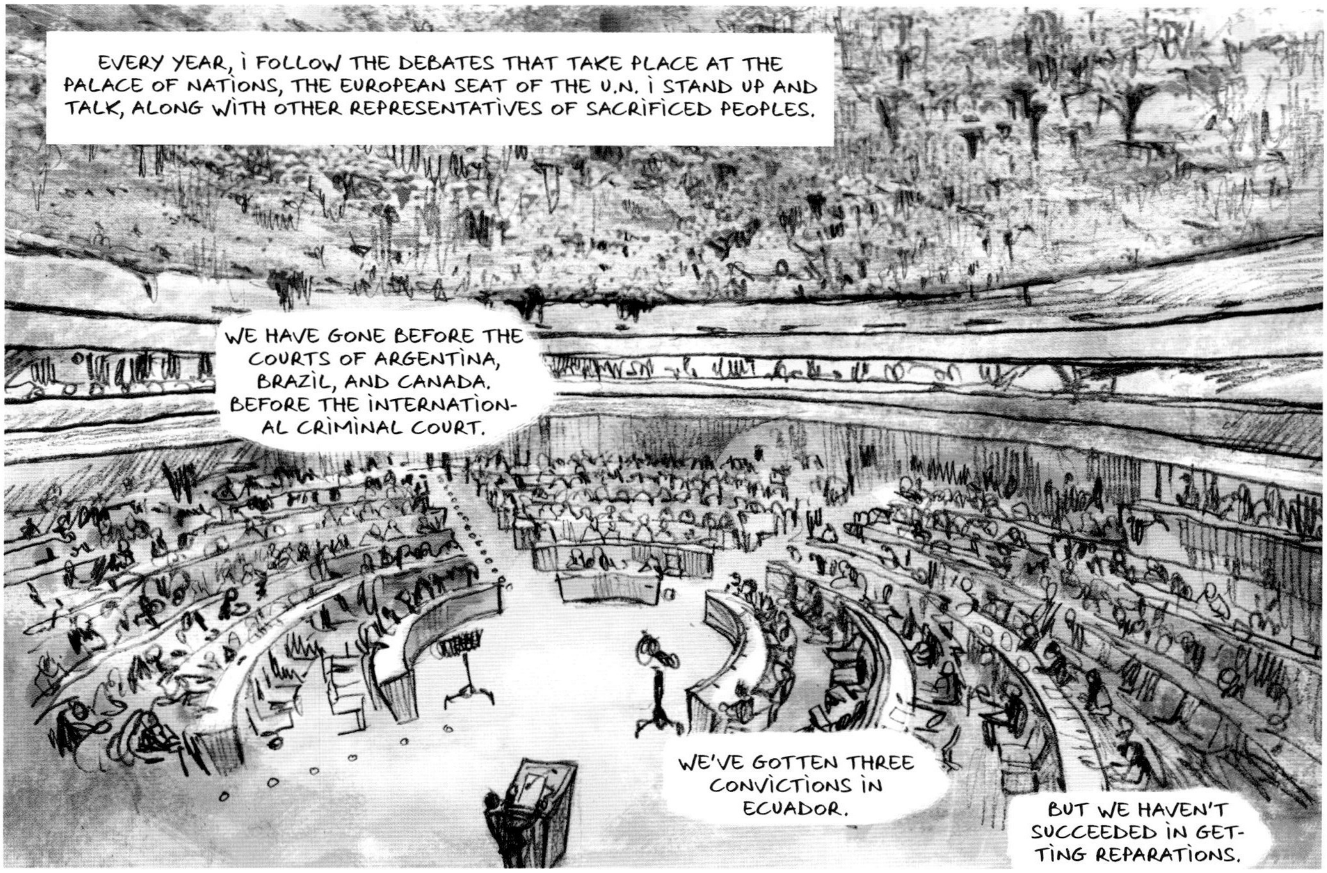
EVERY YEAR, I FOLLOW THE DEBATES THAT TAKE PLACE AT THE PALACE OF NATIONS, THE EUROPEAN SEAT OF THE U.N. I STAND UP AND TALK, ALONG WITH OTHER REPRESENTATIVES OF SACRIFICED PEOPLES.
WE HAVE GONE BEFORE THE COURTS OF ARGENTINA, BRAZIL, AND CANADA. BEFORE THE INTERNATIONAL CRIMINAL COURT.
WE'VE GOTTEN THREE CONVICTIONS IN ECUADOR.
BUT WE HAVEN'T SUCCEEDED IN GETTING REPARATIONS.

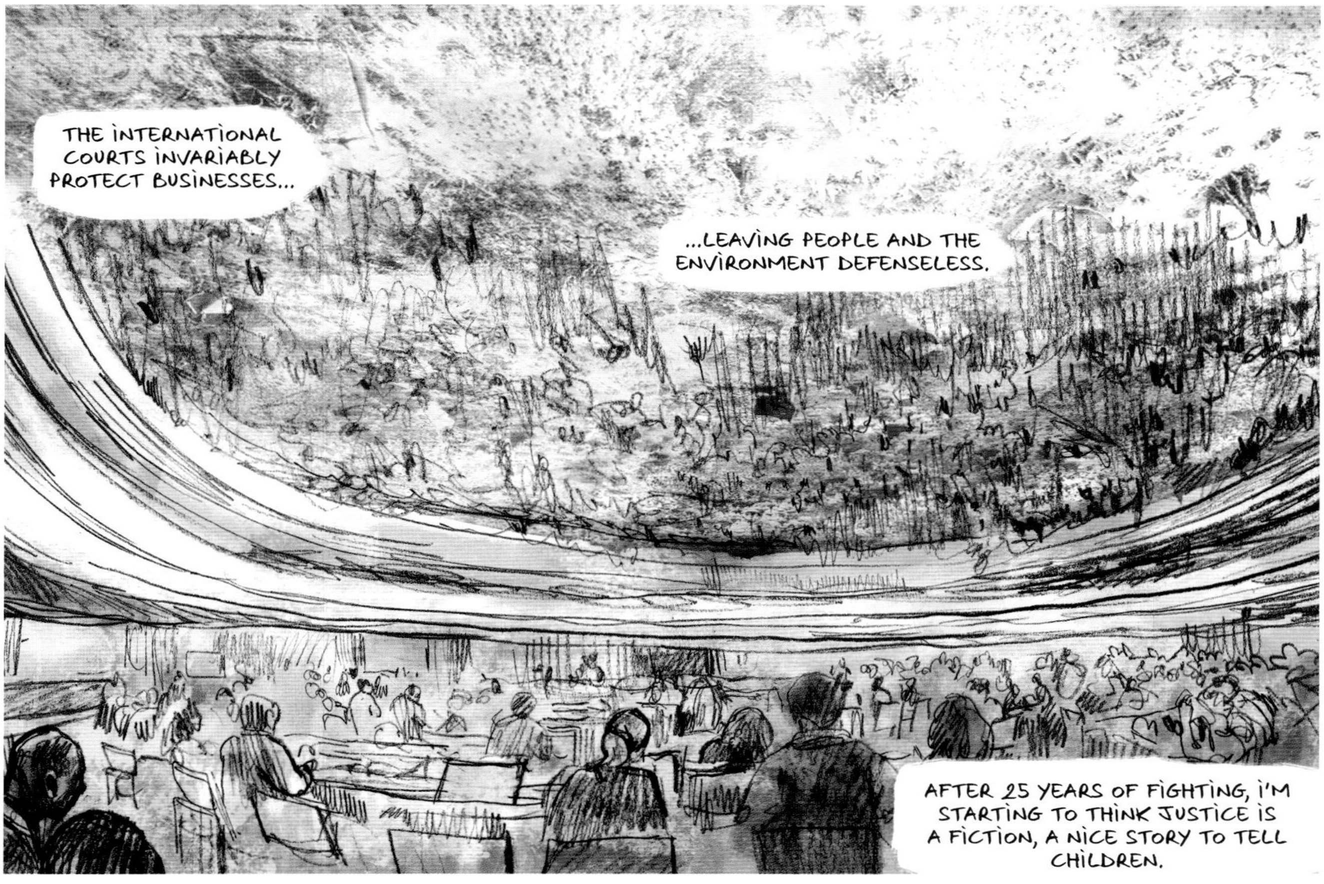
THE INTERNATIONAL COURTS INVARIABLY PROTECT BUSINESSES...
...LEAVING PEOPLE AND THE ENVIRONMENT DEFENSELESS.
AFTER 25 YEARS OF FIGHTING, I'M STARTING TO THINK JUSTICE IS A FICTION, A NICE STORY TO TELL CHILDREN.

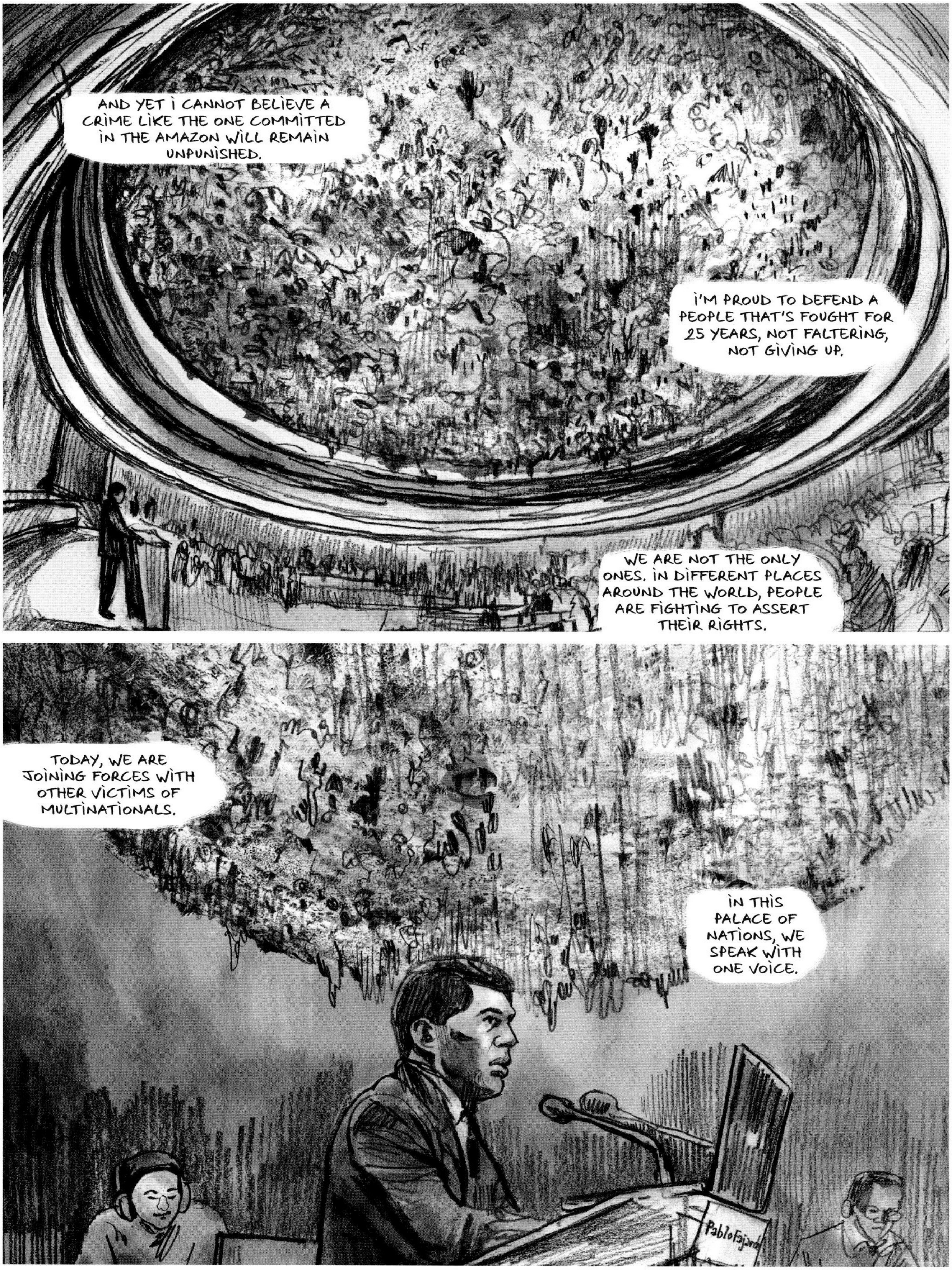
AND YET I CANNOT BELIEVE A CRIME LIKE THE ONE COMMITTED IN THE AMAZON WILL REMAIN UNPUNISHED.
I'M PROUD TO DEFEND A PEOPLE THAT'S FOUGHT FOR 25 YEARS, NOT FALTERING, NOT GIVING UP.
WE ARE NOT THE ONLY ONES. IN DIFFERENT PLACES AROUND THE WORLD, PEOPLE ARE FIGHTING TO ASSERT THEIR RIGHTS.
TODAY, WE ARE JOINING FORCES WITH OTHER VICTIMS OF MULTINATIONALS.
IN THIS PALACE OF NATIONS, WE SPEAK WITH ONE VOICE.

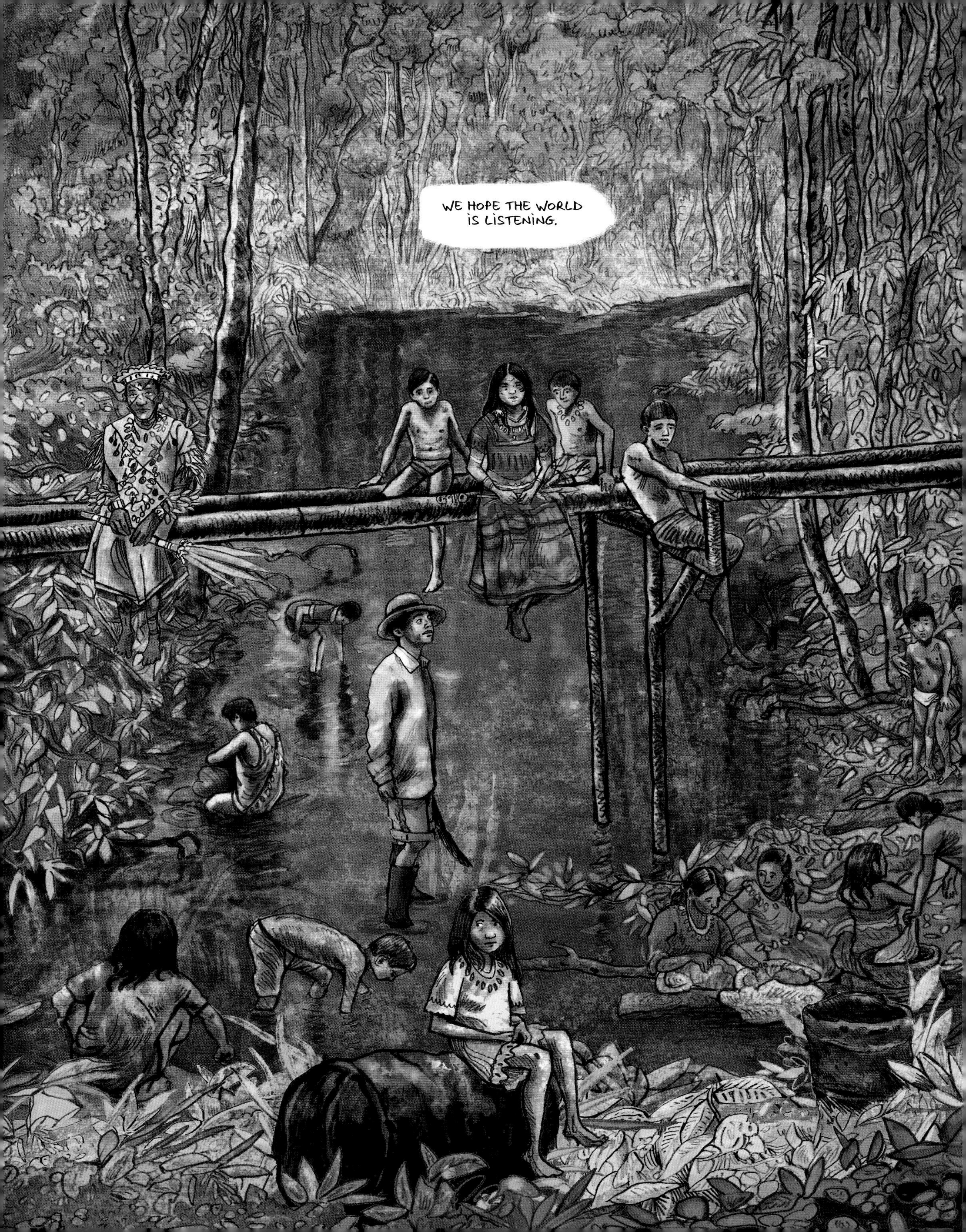
WE HOPE THE WORLD IS LISTENING.

Thanks to Pablo, Donald, Humberto, Julio, Pochito, Mariana, Claudio,
and all the inhabitants of the Lago Agrio region who agreed to tell us their stories;

to Patrick de Saint-Exupéry, the first to see the importance of making this story known;

to Laurent Muller and Anaïs Paris, who trusted us to adapt it into a graphic novel;

to Cédric Liano, behind-the-scenes advisor who, from the storyboard sketches
to the final versions, provided his expert advice and valued friendship;

to Khassatu Ba, intern swiftly promoted to color assistant, as efficient as she is talented;

to Geneviève Garrigos, Pauline David, and the team at Amnesty France for their support.

Sophie and Damien

Thanks to the Unión de Afectados por Texaco, and to the communities of indigenous peoples
of the northern Amazon: Siona, Secoya, Cofán, Kichwa, Shuar, and Waorani.

To the thousands of men and women living in the rural areas of the Amazon,
who have been participating in this common struggle for access to justice for 25 years.
Thank you for never wavering and for continuing to carry this ray of hope for humanity.

To all the NGOs that support and have supported UDAPT's battle
and are working to change the unjust system that runs the world and guarantees impunity
for multinational corporations while leaving the victims of their crimes defenseless.

Thank you for everything that we have accomplished,
and above all for everything that we will continue to accomplish together.

Pablo

An Emblematic Case

Multinational corporations now have branches all over the world, sometimes putting themselves in competition with other nations, especially when their investments or sales exceed the GDP of some countries. These companies no longer know borders, and their subsidiaries are under the jurisdiction of the countries where they are located. Their rapidly expanding operations can be a real headache when it comes to knowing where and how to call them to account for their actions.

Their activities can have lasting and disastrous effects on the lives and rights of thousands of people. This book demonstrates this fact strikingly and shows us the obstacle course that awaits those who seek justice in the face of such behemoths.

Crude perfectly reflects the work that Amnesty International has been doing for several years, documenting and denouncing human rights violations linked to the activities of multinationals: our commitment also aims to support and bring attention to those who dare to defend their rights. It remains very difficult for victims of oil pollution caused by a subsidiary or subcontractor to obtain compensation from the multinational company's headquarters. They usually have to handle the case in their country, where the justice system is often at least flawed, if not corrupt. Furthermore, even where these proceedings are available, they are very lengthy and, during this time, there is a risk that the subsidiary may become insolvent, or even be purchased by another group, which makes it harder to prosecute.

We have known about these injustices for decades. From the Bhopal tragedy in India to the dumping of toxic waste in Côte d'Ivoire and the Niger Delta, the same patterns appear everywhere: a lack of justice, impunity for multinationals, and disregard for victims. The case of Texaco/Chevron shows that the current systems do not hold economic players responsible for human rights, nor are they called to account or compelled to make reparations. Especially since, in the commercial world, arbitration courts—made up of business lawyers instead of judges—can interfere in the judicial process. The Court of Arbitration at The Hague, ruling on a dispute between Chevron and the State of Ecuador, summoned the Ecuadorian government to invalidate the sentence imposed on Chevron in 2011. Though Ecuador has not given in, the threat remains.

Globalization has favored the development of international business activities. There have, however, been some notable developments in recent years. In 2011, the UN adopted guiding principles on business and human rights, which are now used as the international standard.

In June 2014, the UN Human Rights Council, at Ecuador's urging, set up an intergovernmental working group responsible for developing "an international legally binding instrument on transnational corporations and other business enterprises with respect to human rights."

Such an international treaty would make it possible to have businesses considered subjects of international law. They would then be liable for their actions and subject to effective sanctions. The course is set, but not all nations are actively participating in the negotiations. And so the European Union is on the defensive when it should be a driving force.

This process will take time, and it requires ongoing mobilization—mobilization that has already paid off on a different scale, since it has driven France to take a historic step: in March 2017, the country adopted a law concerning the due diligence of parent and contracting companies. This law is the result of the joint effort of a

coalition of NGOs and informed public opinion. It stipulates that large businesses must adopt and publicize environmental impact assessments that include the option for victims or interested parties to bring their case to a judge.

This book is therefore necessary, in that it helps shine light on these realities, because informed opinion can also contribute to change. Additionally, *Crude* is a great tribute to the battle fought by Pablo and the plaintiffs of the Amazon, and with them, all those who are fighting to assert their rights. This is a battle that human rights defenders fight on a daily basis around the world. We must make their work known in order to protect and support them. Defending rights pertaining to land, territory, and the environment is dangerous. When indigenous peoples demand access to their ancestral lands, or ask to be heard regarding the use and exploitation of these lands and their resources, they very often become victims of threats, intimidation, and attacks. Latin America is the world's deadliest region for environmentalists: of all the murders of environmentalists worldwide, about 60 percent are committed there.

This situation is the result of the power imbalance between the defenders, who are often isolated, and the governments, which do not hesitate to sell off land to companies, especially those in the natural resource exploitation sector. If they resist, these defenders leave themselves open to violent acts, most of which go unpunished. And the governments pass these environmental activists off as opponents who threaten the economic development of their countries. However, a country's need for development should not be promoted without regard for the well-being of whole sectors of the population. The role of states is to guarantee equal rights for everyone. It is imperative that governments commit to protecting those who defend the environment, by adopting measures that take indigenous peoples into consideration and are respectful of their particular jurisdictions, as well as national and international regulatory frameworks.

May this inspiring book help to publicize and amplify the battle fought by all those who, like Pablo and the plaintiffs against Chevron, refuse to see their rights sacrificed.

Amnesty International

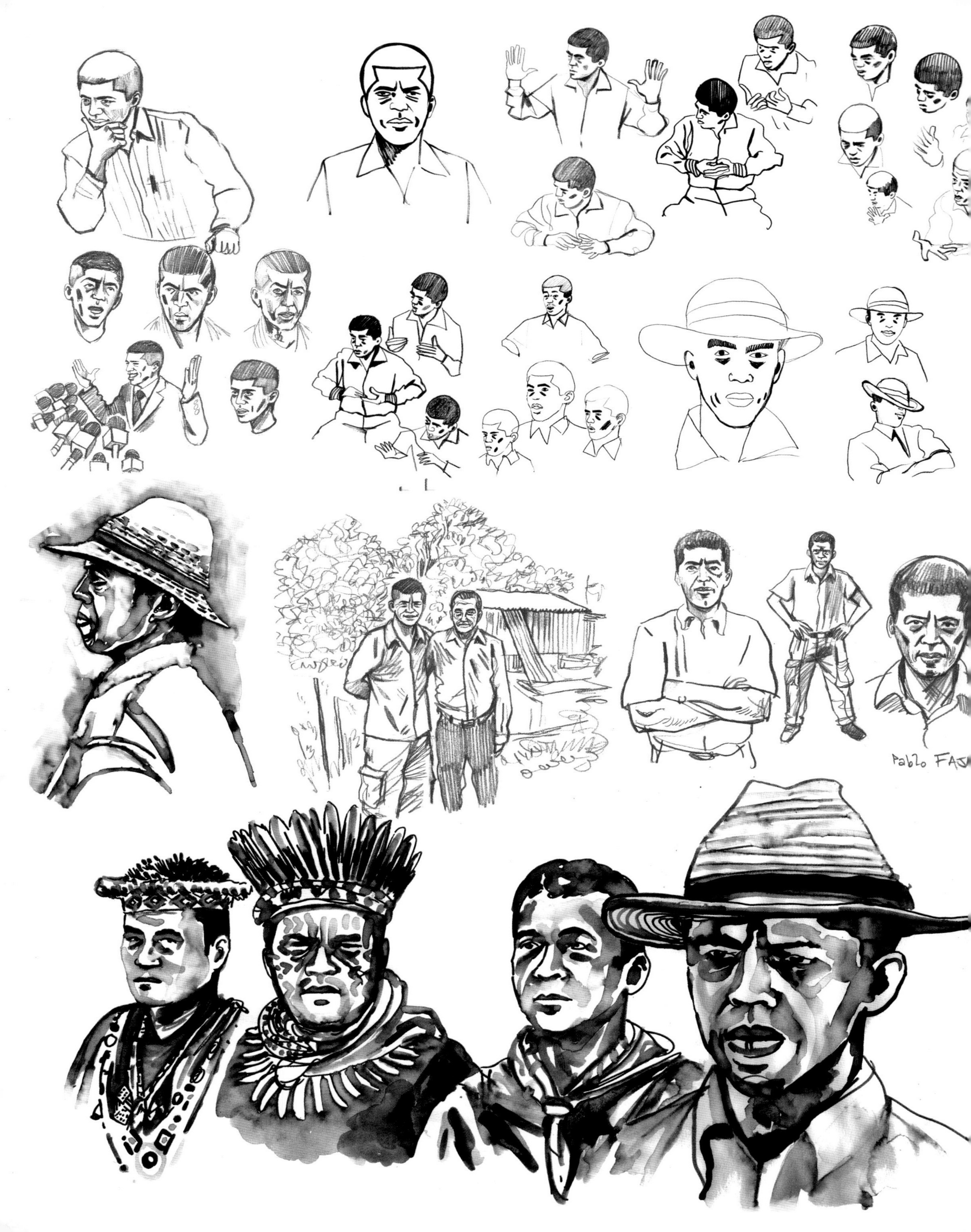
Pablo

Timeline

1964   The government of Ecuador grants a concession to the consortium formed by Gulf and Texaco, so they can look for oil in the Amazon.

1967   Drilling of the first well, Lago Agrio #1.

1969–72   Texaco builds more than 300 oil wells and 18 production stations.

1974   The Ecuadorian government joins the consortium by buying shares from Texaco and Gulf. In 1976, it even becomes the majority shareholder. But all operational decisions remain the exclusive responsibility of Texaco.

1991   *Amazon Crude,* by Judith Kimerling, is published by the National Resources Defense Council.

1992   The concession granted to Texaco expires. The company leaves the country.

1993   The Spanish edition of *Amazon Crude*, titled *Crudo Amazónico*, is published by Ediciones Abya-Yala.

1993   The Union of People Affected by Chevron-Texaco (UDAPT) files a lawsuit against Texaco in New York.

1995   Texaco signs a cleanup contract with the state. However, according to UDAPT, this work is carried out in less than 1 percent of the exploited area.

1998   The government of Ecuador signs an act with Texaco that clears the company of any responsibility and agrees never to take court action against it.

2002   After 9 years of debate on the jurisdiction of the New York court, the case is transferred to Ecuador.

2003   UDAPT files a lawsuit with the court in Lago Agrio, in the Sucumbíos Province. It is the first time that victims of a multinational oil company have brought legal action in their country.

2011   On February 1, Chevron files a lawsuit against "Donziger et al." for organized fraud, and on February 14, Judge Nicolás Zambrano sentences Chevron to a record fine of $9 billion.

2012   UDAPT brings its case to several countries to try to get the Ecuadorian conviction enforced.

2014   Judge Lewis Kaplan of New York condemns UDAPT representatives for fraud. This decision will be confirmed on appeal in 2015.

2015   During an arbitration trial in New York between Chevron and the Ecuadorian State, Judge Alberto Guerra admits to having lied in accusing Pablo Fajardo and Steven Donziger of bribing Judge Zambrano.

2018   On July 10, the Constitutional Court of Ecuador confirms the conviction made by Nicolás Zambrano. It is now irreversible.

Where Are They Now?

Pablo Fajardo
Senior lawyer for the Unión de Afectados y Afectadas por las Operaciones Petroleras de Texaco (UDAPT), he continues to dedicate his life to the Chevron case. With the help of NGOs, universities, and individuals who make donations to UDAPT, he travels all over the world to advocate for the Afectados and try to uphold the rights of humans and the environment.

Julio Prieto
Prieto has worked actively with UDAPT since 2005 and is now an environmental lawyer. He holds a law degree and a master's degree in environmental management from the Yale School of Forestry and Environmental Studies.

Humberto Piaguaje
Historic leader of the Secoya people, Piaguaje has been involved with the legal battle against Chevron from the beginning. He was the coordinator of UDAPT from 2012 to 2018 and lives in Lago Agrio with his wife and children.

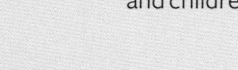

Steven Donziger
After he won the case against Chevron in 2011, Chevron countersued. In 2014 Donziger was convicted of racketeering in a U.S. federal court. He was placed under house arrest in 2019, and in August 2020, he was stripped of his law license.

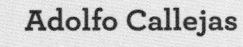

Adolfo Callejas
Since 1974, this Ecuadorian lawyer has devoted his entire career to Chevron. Today, he still works on behalf of the multinational corporation.

Alberto Guerra

Formerly a judge in Ecuador, Guerra was removed from office in 2008 following charges of corruption in a drug-trafficking case. Guerra has since moved to Florida and, according to UDAPT, has been paid more than $2 million dollars by the multinational corporation for serving as a key witness in Chevron's legal battles.

Joe Berlinger

He spent $1.3 million in legal fees to fight the Chevron subpoena demanding outtakes from *Crude*. After losing his case, Berlinger was forced to turn over the footage, as well as his personal computer, to the U.S. courts. He has said that he would have to arm himself with a sizable legal war chest if he ever were to make another film that is critical of a large corporation.

Nicolás Zambrano

Removed from office as judge in 2012 on suspicion of fraud in a drug-trafficking case, he now works in the private sector and hopes to return to the judiciary.

Mariana Jiménez

Born in the south of Ecuador, she is one of the "colonists" who settled in the Lago Agrio region. She has lived there for fifty years and is very active within UDAPT.

Judith Kimerling

She is currently Professor of Environmental Law and Policy at Queens College, CUNY. After writing the book that opened the door for the Chevron case, Kimerling has continued to do groundbreaking research on oil operations in the Amazon as well as advocate for the Baihuaeri Waorani of Bameno, who are working to stop the expansion of oil operations in their territory and defend their human rights.

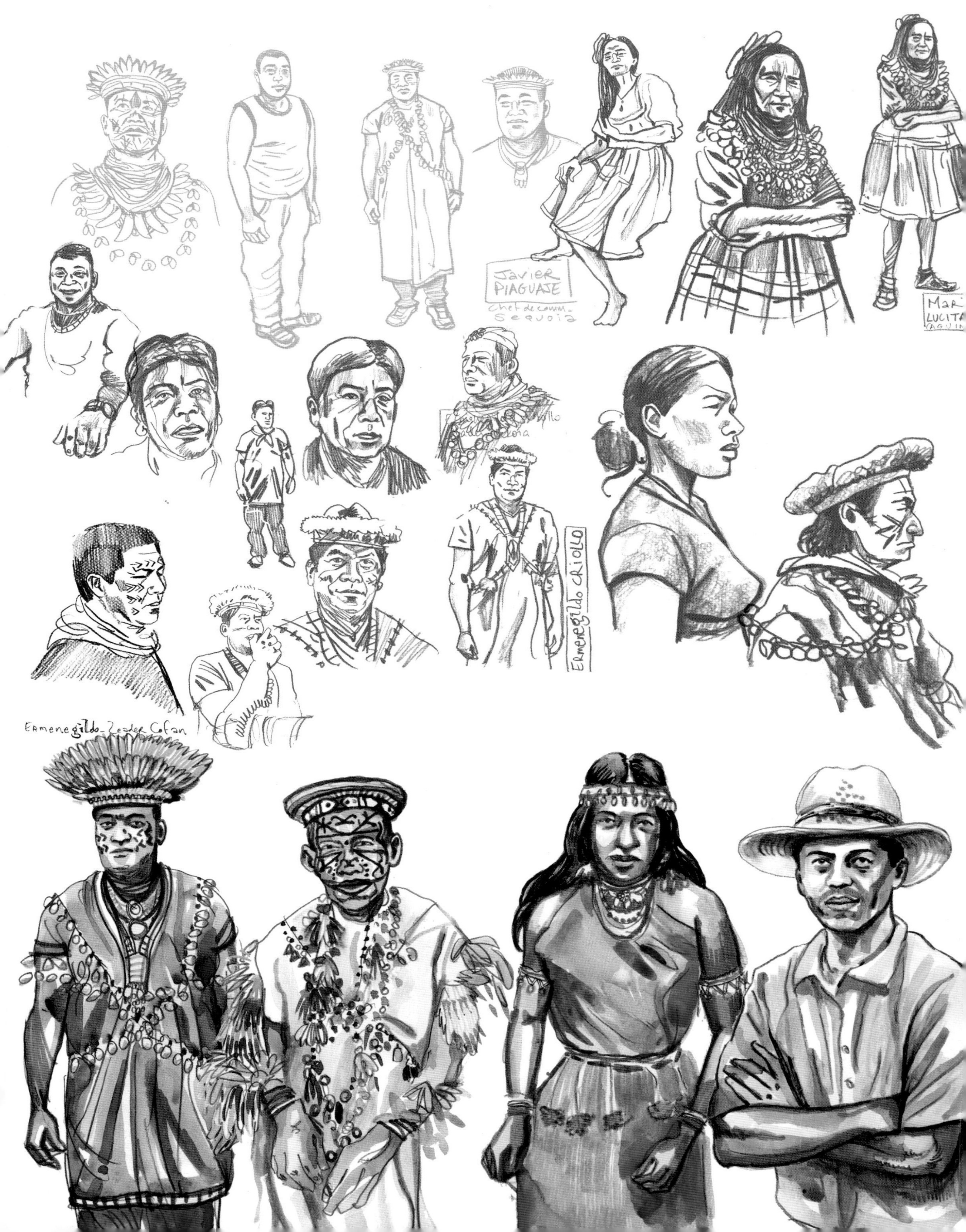
Javier
PIAGUAJE
Chef de comm.
Sequoia
Mar
LUCITA
Ermenegildo CRIOLLO
Ermenegildo_Leader Cofan

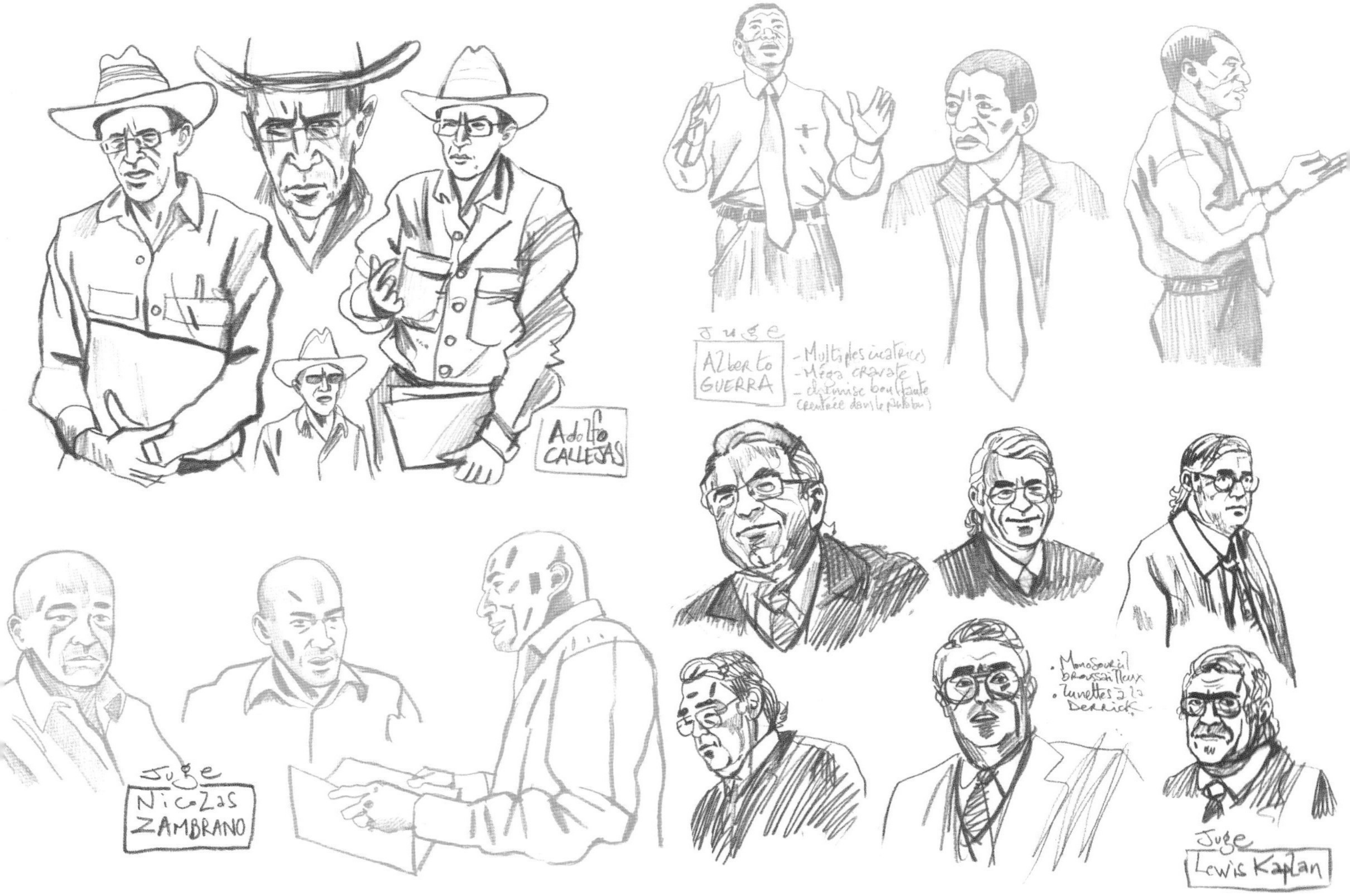
Adolfo CALLEJAS
Juge
Alberto GUERRA
- Multiples cicatrices
- Méga cravate
- chemise bouffante (rentrée dans le pantalon)
Juge
NicoLas ZAMBRANO
• Monosourcil broussailleux
• Lunettes à la Derrick
Juge
Lewis Kaplan

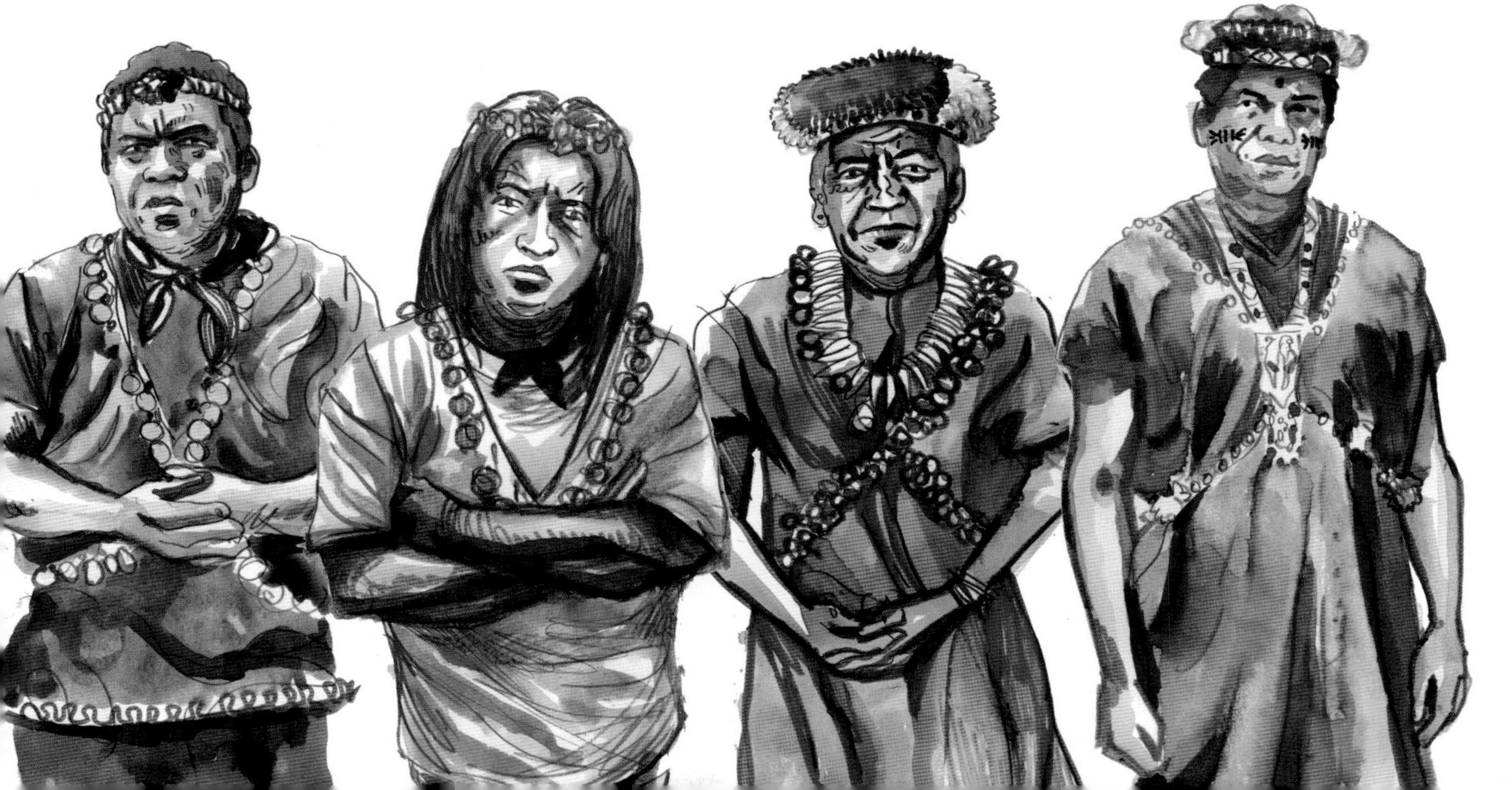

A Global Suit

The case between UDAPT and Chevron continues around the world today. The plaintiffs have gone to several foreign courts in order to have the Ecuadorian judgment recognized and Chevron's assets seized. For its part, the company criminalizes the plaintiffs and attacks the Government of Ecuador.

CANADA

In 2012, UDAPT filed a request to have the Ecuadorian conviction enforced in Canada. The Supreme Court declared that it had jurisdiction to hear the case. The plaintiffs, however, were not able to seize the assets of Chevron Canada, since the judges found that the subsidiary is a separate business from the parent company. The Supreme Court must now give a substantive ruling.

UNITED STATES

On February 1, 2011, Chevron filed a complaint against "Donziger et al." for fraud under the RICO Act, which was designed to combat drug trafficking and organized crime. In March 2014, Judge Lewis Kaplan sided with Chevron, judging that UDAPT representatives used Mafia methods to extort money from a very rich company. The decision, which also prohibits them from enforcing the sentence in the United States, was upheld on appeal in 2015.

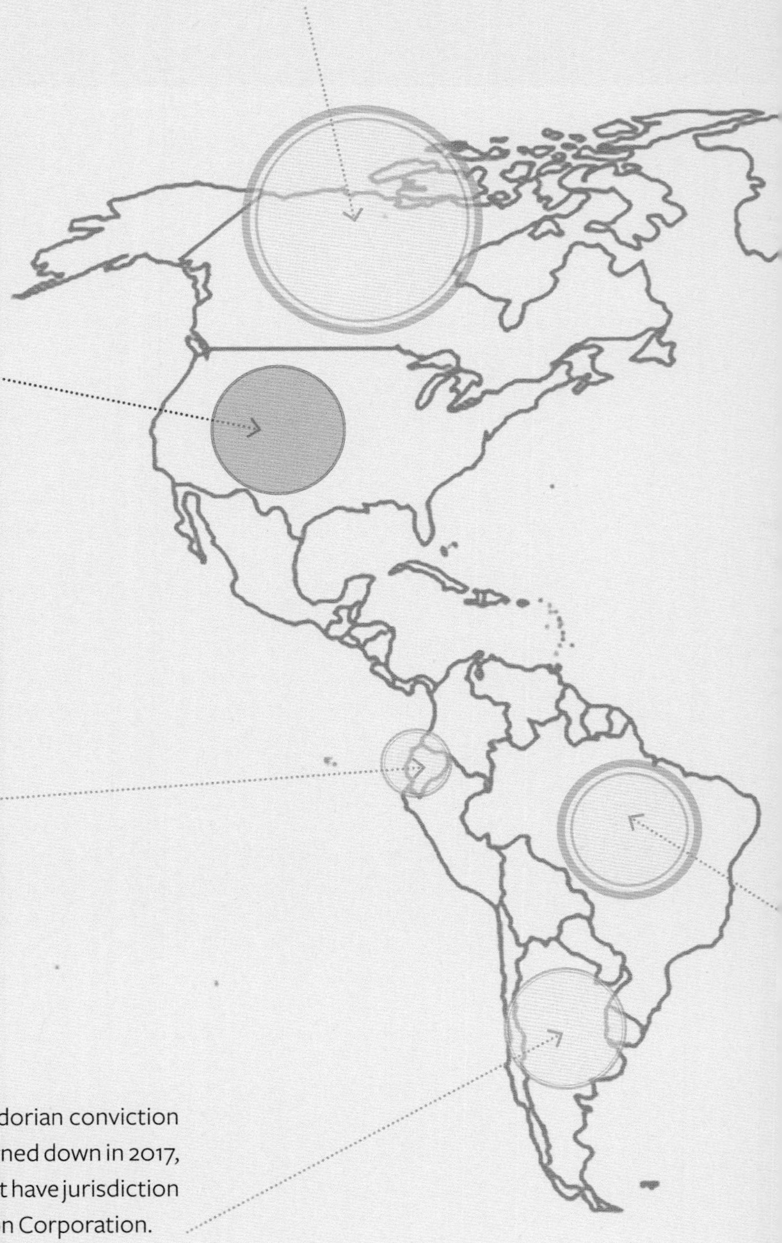

ECUADOR

On February 14, 2011, the court in Lago Agrio sentenced Chevron to a record fine of $9 billion. This decision was upheld on appeal in January 2012, then by the Supreme Court in November 2013. On June 27, 2018, the conviction was confirmed by the Constitutional Court, the country's highest court. It is now irreversible.

ARGENTINA

In 2012, UDAPT lodged an appeal to have the Ecuadorian conviction enforced on Argentine territory. Its request was turned down in 2017, as the Argentine justice system determined it did not have jurisdiction to judge a case against the parent company, Chevron Corporation.

THE HAGUE - INTERNATIONAL CRIMINAL COURT

In 2014, UDAPT filed a lawsuit before the International Criminal Court in The Hague against John Watson, the CEO of Chevron, accusing him of crimes against humanity for his refusal to carry out the cleanup. This appeal was dismissed in March 2015.

THE HAGUE - PERMANENT COURT OF ARBITRATION

Chevron has made three requests for arbitration against the Ecuadorian State. These appeals are not strictly about UDAPT but are instead related to the government, which they have accused of not respecting the agreements securing Chevron's investments in the country. On August 30, 2018, the arbitrators ruled in the company's favor, ordering Ecuador to reimburse Chevron for the legal costs it incurred to defend itself. They also called for the government to overturn the conviction made in 2011 by Judge Zambrano. The State cannot honor this request without violating its own Constitution, which separates executive and legislative powers.

UN HUMAN RIGHTS COUNCIL

Since 2014, UDAPT has regularly appeared before the Human Rights Council in Geneva to denounce Chevron's crime and to advocate for the adoption of a treaty that would compel multinational corporations to respect the rights of peoples.

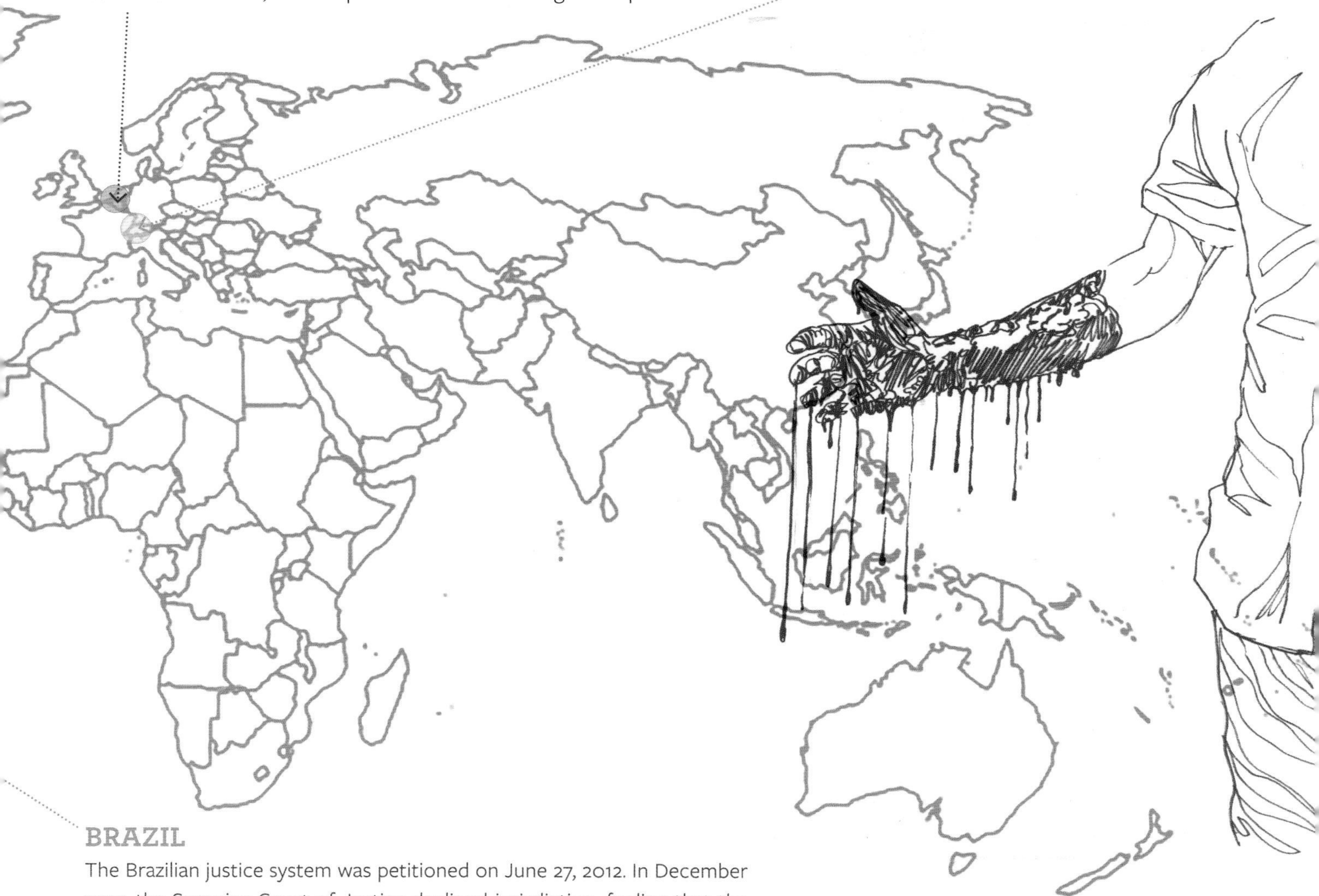

BRAZIL

The Brazilian justice system was petitioned on June 27, 2012. In December 2017, the Superior Court of Justice declined jurisdiction, finding that the Brazilian courts are unable to impose a sentence on Chevron Corporation.

Cases brought by UDAPT

Cases brought by Chevron

Resources

Barrett, Paul M. *Law of the Jungle: The $19 Billion Legal Battle over Oil in the Rain Forest and the Lawyer Who'd Stop at Nothing to Win*. New York: Broadway Books, 2014.

Berlinger, Joe, dir. *Crude*. Entendre Films; RadicalMedia; Red Envelope Entertainment; Third Eye Motion Picture Company, 2009.

"Chevron Wins Ecuador Rainforest 'Oil Dumping' Case." BBC, September 7, 2018. https://www.bbc.com/news/world-latin-america-45455984.

Coronel Vargas, Gabriela, William W. Au, and Alberto Izzotti. "Public Health Issues from Crude-Oil Production in the Ecuadorian Amazon Territories." *Science of the Total Environment* 719, no. 1 (June 2020). https://doi.org/10.1016/j.scitotenv.2019.134647.

Goldhaber, Michael D. *Crude Awakening: Chevron in Ecuador*. New York: RosettaBooks, 2014.

Hecht, Susanna B., and Alexander Cockburn. *The Fate of the Forest: Developers, Destroyers, and Defenders of the Amazon*. Chicago: University of Chicago Press, 2010.

Kimerling, Judith. *Amazon Crude*. New York: Natural Resources Defense Council, 1991.

Langewiesche, William. "Jungle Law." *Vanity Fair,* April 3, 2007. https://www.vanityfair.com/news/2007/05/texaco200705.

Lerner, Sharon. "How the Environmental Lawyer Who Won a Massive Judgement Against Chevron Lost Everything." *Intercept*, January 29, 2020. https://theintercept.com/2020/01/29/chevron-ecuador-lawsuit-steven-donziger.

Long, Gideon. "Ecuador's Indigenous People Under Threat from Oil Drilling." *Financial Times*, December 4, 2019. https://www.ft.com/content/8e1acf14-e467-11e9-b8e0-026e07cbe5b4.

Orellana López, Aldo. "Chevron vs Ecuador: International Arbitration and Corporate Impunity." openDemocracy, March 27, 2019. https://www.opendemocracy.net/en/democraciaabierta/chevron-vs-ecuador-international-arbitration-and-corporate-impunity.

"Pablo Fajardo Mendoza and Luis Yanza: 2008 Goldman Prize Recipient, South and Central America." Goldman Environmental Prize, 2008. https://www.goldmanprize.org/recipient/pablo-fajardo-mendoza-luis-yanza/.

Parloff, Roger. "Attorney Who Took Chevron to Court for $18 Billion Suspended by Amazon Defense Front." *Fortune*, July 31, 2016. https://fortune.com/2016/07/31/adf-suspends-pablo-fajardo-attorney-who-took-chevron-to-court-for-18-billion.

Romero, Simon, and Clifford Krauss. "In Ecuador, Resentment of an Oil Company Oozes." *New York Times*, May 14, 2009. https://www.nytimes.com/2009/05/15/business/global/15chevron.html.

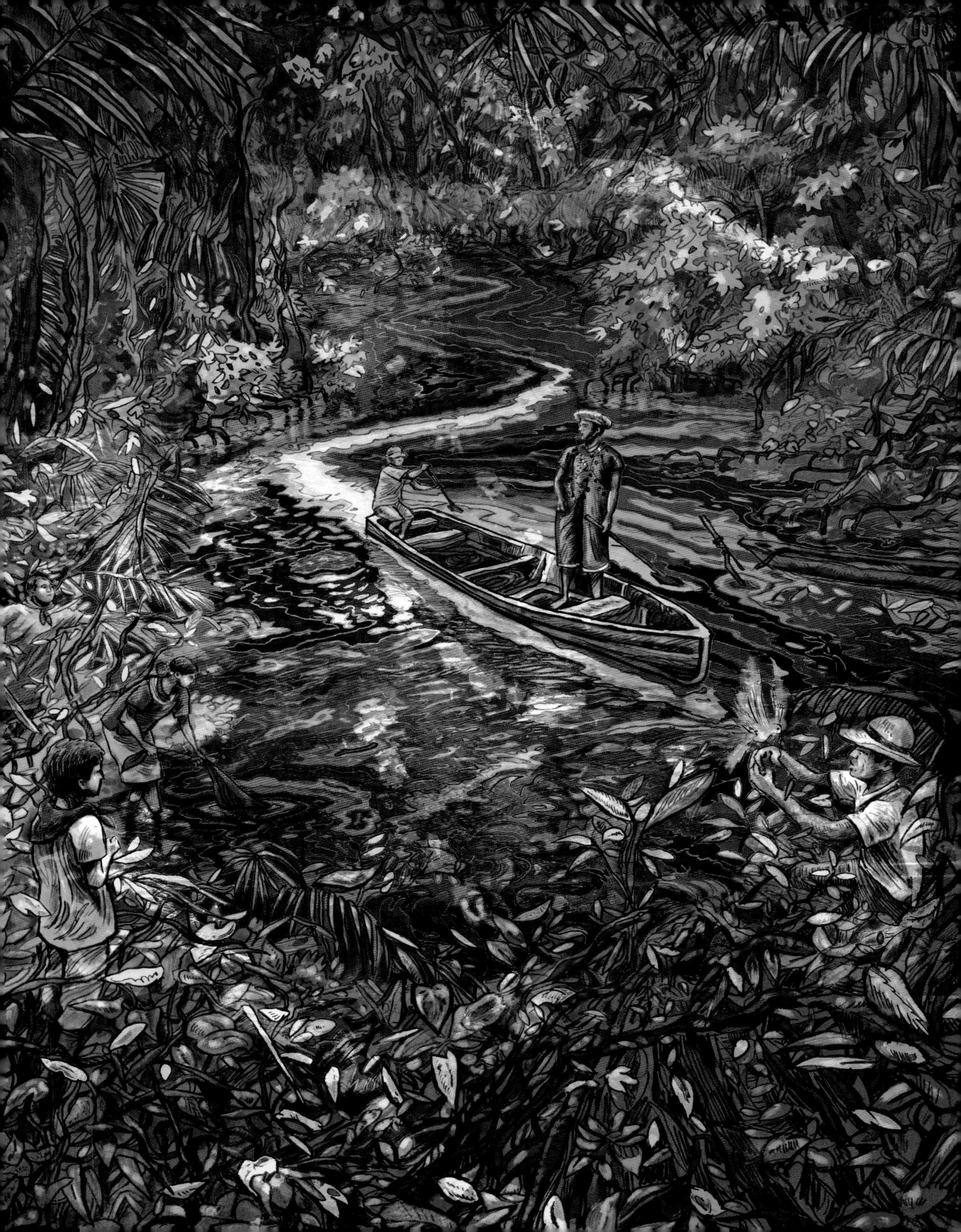

Library of Congress Cataloging-in-Publication Data

Names: Fajardo, Pablo 1972–, author.
Title: Crude: a memoir / story, Pablo Fajardo; script, Sophie Tardy-Joubert; drawing and color, Damien Roudeau; translated by Hannah Chute.
Description: University Park, Pennsylvania: The Pennsylvania State University Press / Graphic Mundi, [2021] | Includes bibliographical references.
Summary: "A graphic novel exploring Texaco's involvement in the Amazon, as well as the ensuing legal battles between the oil company, the Ecuadorian government, and the region's inhabitants, from the perspective of Ecuadorian lawyer and activist Pablo Fajardo"—Provided by publisher.
Identifiers: LCCN 2020054793 | ISBN 9780271088068 (hardback ; alk. paper)
Subjects: Fajardo, Pablo, 1972—-Comic books, strips, etc. | Texaco, Inc.—Comic books, strips, etc. | Chevron Corporation (2005)—Comic books, strips, etc. | Petroleum industry and trade—Environmental aspects—Ecuador—Oriente—Comic books, strips, etc. | Petroleum waste—Environmental aspects—Ecuador—Oriente—Comic books, strips, etc. | Oil spills—Environmental aspects—Ecuador—Oriente—Comic books, strips, etc. | Liability for oil pollution damages—Ecuador—Comic books, strips, etc. | Indians of South America—Ecuador—Oriente—Social conditions—Comic books, strips, etc.
Classification: LCC TD195.P4
LC record available at https://lccn.loc.gov/2020054793

Editorial manager: Laurent Muller
Editorial assistant: Anaïs Paris
Graphic designer: Damien Roudeau & Constance Rossignol
Color assistant: Khassatu Ba
Typographer: Thibaut Chignaguet
Copyediting: Ève Sorin
Production: Sarah Joulia & Marie Baird-Smith

Printed in China
Published by The Pennsylvania State University Press,
University Park, PA 16802-1003

Translated by Hannah Chute

Originally published as *Texaco*
© Editions Les Arènes, Paris 2019

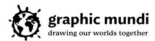

graphic mundi
drawing our worlds together

Graphic Mundi is an imprint of
The Pennsylvania State University Press.

The Pennsylvania State University Press is a member of the Association of University Presses.

It is the policy of The Pennsylvania State University Press to use acid-free paper. Publications on uncoated stock satisfy the minimum requirements of American National Standard for Information Sciences—Permanence of Paper for Printed Library Material, ANSI Z39.48–1992.

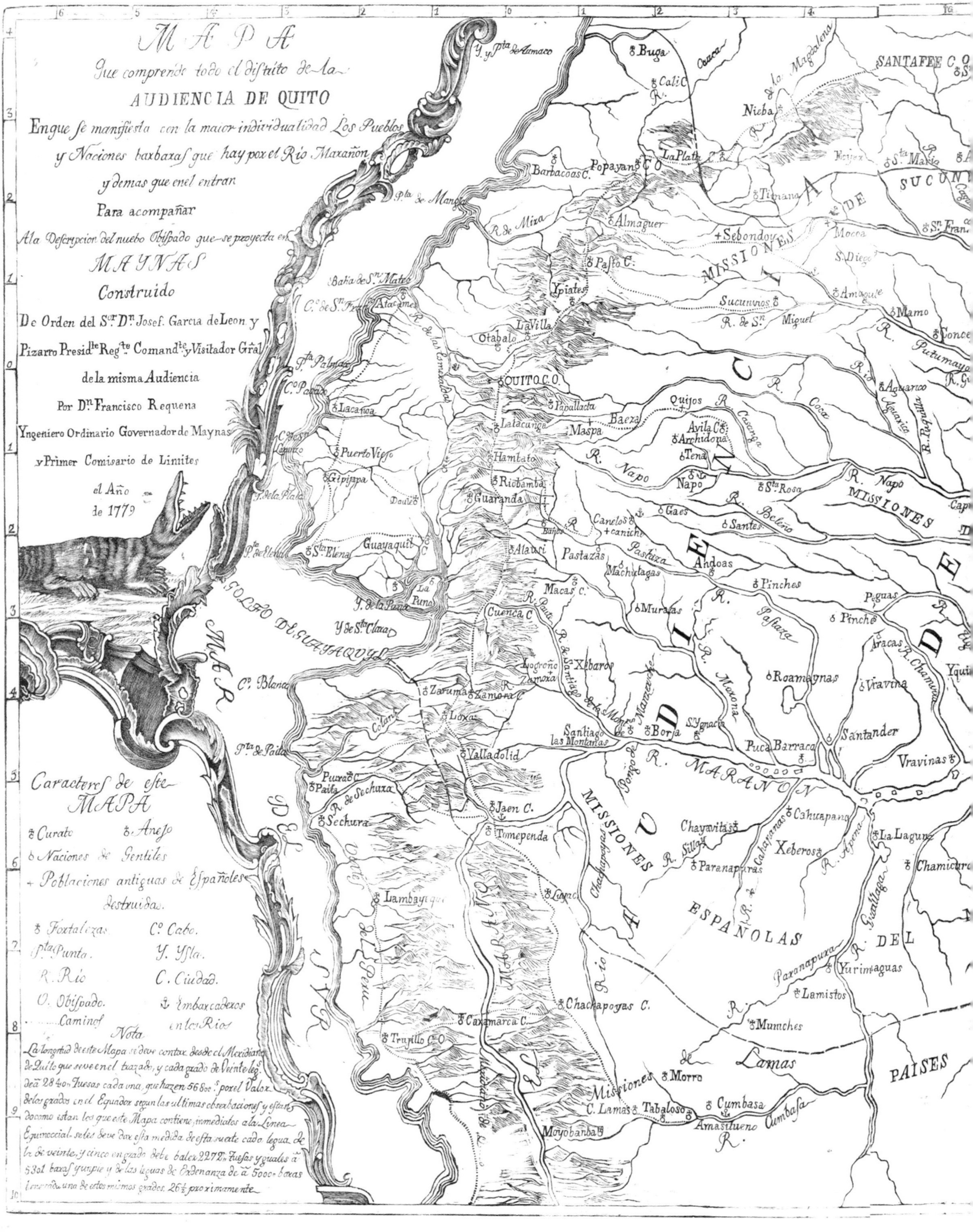
MAPA
Que comprende todo el distrito de la
AUDIENCIA DE QUITO
En que se manifiesta con la maior individualidad Los Pueblos
y Naciones barbaras que hay por el Rio Marañon
y demas que en el entran
Para acompañar
A la Descripcion del nuebo Obispado que se proyecta en
MAYNAS
Construido
De Orden del Sor. Dn. Josef. Garcia de Leon y
Pizarro Presidte. Regte. Comandte. y Visitador Gral.
de la misma Audiencia
Por Dn. Francisco Requena
Yngeniero Ordinario Governador de Maynas
y Primer Comisario de Limites
el Año
de 1779
Caracteres de este
MAPA
Curato
Anejo
Naciones de Gentiles
+ Poblaciones antiguas de Españoles destruidas.
Fortalezas.
Co. Cabo.
Pta. Punta.
Y. Ysla.
R. Rio.
C. Ciudad.
O. Obispado.
Embarcaderos en los Rios
Caminos
Nota
La longitud de este Mapa se deve contar desde el Meridiano de Quito que sirve en el trazado, y cada grado de Veinte legs. de a 2840 Tuesas cada una, que hazen 5680 Ts. por el Valor de los grados en el Equador segun las ultimas observaciones y estando como estan los que este Mapa contiene inmediatos a la Linea Equinoccial se les deve dar esta medida de esta suerte cada legua de las de veinte, y cinco en grado debe baler 2272 Tuesas y guales a 5301 baras y un pie y de las leguas de Ordenanza de a 5000 baras tiene cada una de estos mismos grados 26½ proximamente.
MAR DEL SUR
GOLFO DE GUAYAQUIL
AUDIENCIA
MISSIONES
MISSIONES ESPAÑOLAS
MISSIONES DE SUCUMBIOS
PAISES DEL
R. MARAÑON
QUITO C.O.
Cuenca C.
Loxa
Guayaquil C.
Popayan C.O.
SANTAFEE C.O.
Jaen C.
Chachapoyas C.
Caxamarca C.
Trujillo C.O.
Moyobamba
Borja
Lamas
R. Napo
R. Putumayo
R. Pastaza
Riobamba
Latacunga
Otabalo
Pasto C.
Barbacoas C.
Buga
Cali C.
La Plata C.
Nieba
Timana
Almaguer
Ypiales
Hambato
Guaranda
Alausi
Zaruma
Zamora C.
Valladolid
Piura C.
Paita
Sechura
Lambayeque
Tomependa
Santiago de las Montañas
Andoas
Pinches
Santander
Vravinas
La Laguna
Chamicuro
Yurimaguas
Lamistos
Munches
Morro
Tabaloso
Cumbasa
Xeberos
Cahuapana
Chayavitas
Paranapuras
Roamaynas
Muratas
Machutagas
Archidona
Tena
Avila C.
Napo
Sta. Rosa
Baeza
Quijos
Maspa
Papallacta
Lacanoa
Puerto Viejo
Gipijapa
Daule
Sta. Elena
La Puna
Y. de la Puna
Y. de Sta. Clara
Co. Blanco
Pta. de Paita
Pta. de Sta. Elena
Pta. de Manglares
Pta. de Palmar
Co. Pasado
Bahia de Sn. Mateo
Atacames
Y. y Pta. de Tumaco
R. de Mira
R. de las Esmeraldas
Sucumbios
R. de Sn. Miguel
Sebondoy
Mocoa
S. Diego
Amaguie
Mamo
R. Aguarico
R. Coca
R. Beleño
Gaes
Santes
Canelos
R. Chambira
Aracas
Vravina
Pegnas
Puca Barraca
R. Morona
R. Santiago
Logroño
Xebaros
Pongo de Manseriche
R. Paute
Macas C.
Banos
Pastazas
R. Sillay
R. Cahuapanas
R. Apena
R. Guallaga
R. Paranapura
Rio Chachapoyas
R. Mayo
Colan
Valle del Peru
Lamas
C. Lamas
Amasifueno
R. Cumbasa
R. de la Magdalena
Cauca